Self-Validating Numerics
for
Function Space Problems

Notes and Reports
in
Computer Science and Applied Mathematics

Editor
Werner Rheinboldt
University of Pittsburgh

1. JOHN R. BOURNE. Laboratory Minicomputing

2. CARL TROPPER. Local Computer Network Technologies

3. KENDALL PRESTON, JR. AND LEONARD UHR (Editors). Multicomputers and Image Processing: Algorithms and Programs

4. STEPHEN S. LAVENBERG (Editor). Computer Performance Modeling Handbook

5. R. MICHAEL HORD. Digital Image Processing of Remotely Sensed Data

6. SAKTI P. GHOSH, Y. KAMBAYASHI, AND W. LIPSKI (Editors). Data Base File Organization: Theory and Applications of the Consecutive Retrieval Property

7. ULRICH W. KULISCH AND WILLARD L. MIRANKER (Editors). A New Approach to Scientific Computation

8. JACOB BECK, BARBARA HOPE, AND AZRIEL ROSENFELD (Editors). Human and Machine Vision

9. EDGAR W. KAUCHER AND WILLARD L. MIRANKER. Self-Validating Numerics for Function Space Problems: Computation with Guarantees for Differential and Integral Equations

Self-Validating Numerics for Function Space Problems

Computation with Guarantees for Differential and Integral Equations

Edgar W. Kaucher

Institute for Applied Mathematics
University of Karlsruhe
Karlsruhe, West Germany

Willard L. Miranker

Mathematical Sciences Department
IBM Thomas J. Watson Research Center
Yorktown Heights, New York

1984

ACADEMIC PRESS, INC.
(Harcourt Brace Jovanovich, Publishers)

Orlando San Diego San Francisco New York London
Toronto Montreal Sydney Tokyo São Paulo

ACADEMIC PRESS, INC.
Orlando, Florida 32887

United Kingdom Edition published by
ACADEMIC PRESS, INC. (LONDON) LTD.
24/28 Oval Road, London NW1 7DX

Library of Congress Cataloging in Publication Data

Kaucher, Edgar W.
    Self-validating numerics for function space problems.

    (Notes and reports in computer science and applied
mathematics)
    Bibliography: p.
    1. Function spaces. 2. Numerical analysis.
I. Miranker, Willard L. II. Title. III. Series.
QA323.K38   1984   515.7'3   84-3071
ISBN 0-12-402020-8 (alk. paper)

PRINTED IN THE UNITED STATES OF AMERICA

84 85 86 87   9 8 7 6 5 4 3 2 1

To Marianna and to Phyllis

CONTENTS

PREFACE

Scientific computation has always contributed to the development of the computer. In the past, the demands made by scientific and technical users have dominated this development. The contemporary digital computer and its associated systems are truly impressive, especially compared to predecessors on a scale of decades. We have seen enormous increases in speed and memory size as well as qualitative changes such as the introduction of floating-point, source language development, and the use of graphics.

Computers are now so powerful and so available that the practice of using them in scientific computation tends to conceal important shortcomings. The danger of imprinting these shortcomings is a real one.

(a) Numerical analysis is the bridge between the body of mathematical methodology and computation. Numerical analysis has generated the data types of scientific computation (floating-point numbers, vectors, and matrices over these, complex versions as well as intervals of all of these). The sheer bulk of computation that is performed endows a sense of universality to this collection of data types and the arithmetic operations associated with them. This viewpoint amounts to a limitation of approach that unnecessarily curtails the applicability of both mathematics in computation and of the digital computation process itself.

(b) Its advanced scientific computational capability notwithstanding, the contemporary computer is nevertheless an experimental tool rather than a precise instrument. The significance and accuracy of the results of a computation are not specified by the computer. The user is required to accept results with a feeling rather than a certain knowledge of their meaning. The sophisticated user is often obliged to assess results with side calculations, reruns, and the like. This costly process is itself imprecise, and its complexity proscribes its use by the naïve computer user.

Here we develop computational methods for solving function space problems such as differential equations and integral equations. These methods address the two shortcomings just described. The function space problems are made capable of a direct encoding into the digital computer in terms of data types for funcions. The operations of arithmetic, as well as certain infinitesimal operations for functions (such as integration), are made available as basic computer operations for the function data types. This collection of operations comprise what we call ultra-arithmetic. The operations are defined by means of projection or rounding operators taking the function space into a finite dimensional space corresponding to the function data types. The structure comprised of both the function data types and ultra-arithmetic is called a functoid. With ultra-arithmetic, problems assocaited with functions become solvable on computers just as algebraic problems are solvable on traditional computers.

Validation techniques are introduced so that the computation itself supplies guarantees concerning the solution of the function space problem being computed. Proceeding solely from the problem data itself, existence, uniqueness, and bounds for the solution that is sought are provided computationally. The method for doing this is to use an interval ultra-arithmetic for simulating a fixed point iteration. The computation produces a set of functions inside of which the solution of the function space problem (and not merely some discretized version) is guaranteed to exist. The set of functions is specified by its boundaries, the latter being composed of computer representable functions, that is, by function data types.

The quality of the bounds provided by the validation technique can usually be made to be very high (of least-significant-bit accuracy). This is accomplished by employing methods of iterative residual correction in function spaces that we develop.

To exploit these capabilities of computation in a functoid, a computer's source language must be augmented to accommodate the function data type and the augmented set of the operations that comprise ultra-arithmetic. We discuss these source language questions.

We treat the actual implementation of the methodology introduced and developed here. A detailed simulation of it has been carried out. Results of self-validated computations in a functoid for a number of function space problems are shown.

We expect further development of these methods to occur. In turn, emphasis in numerical analysis will be brought into the function space approach to numerical problems (such as Galerkin methods, spectral methods, and the tau-method). Following this, we expect that computers themselves will evolve to provide both function data types and ultra-arithmetic as a basic set of operations for scientific computation. Techniques of self-validation will be improved and made efficient. Along with the development of computers that perform ultra-arithmetic will come an associated systems and source language development. In this manner scientific computation will again provide a significant impulse to the continued evolution of the computer.

ACKNOWLEDGMENTS

The authors are grateful to U. Kulisch, Director of the Institute for Applied Mathematics at the University of Karlsruhe, for his interest in this work and for his support and encouragement of its preparation. The authors are grateful to S. Winograd, Director of the Mathematical Sciences Department of the IBM Thomas J. Watson Research Center, for his interest and support as well.

We are grateful to D. Cordes and T. Kölmel, who performed the numerical computations reported on in Chapters 5 and 7.

We thank J. Genzano and B. J. White for the preparation of the manuscript. This manuscript was typeset at the IBM Thomas J. Watson Research Center on the Autologic APS5 Phototypesetter. YFL was used to format the file, and XEDIT on a 3277 terminal running VM/SP was used to create the file. Figures were prepared by the Graphics Department at the IBM Thomas J. Watson Research Center.

The work of E. W. Kaucher was supported by the IBM Thomas J. Watson Research Center during August 1982. The work of W. L. Miranker was partially supported by the Alexander von Humboldt Foundation during a stay at the University of Karlsruhe, 1982–1983.

Chapter 1

INTRODUCTION

The requirements of scientific and engineering computation have typical-ly exceeded the capability of computer methodology in the quantitative sense. The capacity afforded by a computer of increased performance or by a method of increased scope is quickly consumed by an increase in scale of the problems to be solved numerically. However, computer methodology has a qualitative limitation for scientific and engineering computation as well. This is a limitation of accuracy and significance of the results that are produced. The growing scale of contemporary computation has only made this qualitative limitation all the more evi-dent and all the more critical.

The use of a computer as an experimental tool rather than as a precise instrument requires the user to accept results with a feeling for rather than a certain knowledge of their meaning. This is a costly process since it requires the expenditure of human and machine resources in attempting to assess results with checks or reruns and the like. Today the costs of even slight unreliability in computation in, say, a robotic assembly line, a power plant, or a weapons system add force to the need to address this qualitative limitation of accuracy and significance in computation.

Changing computing methodology so that the computer delivers validated results will require a computational environment comprising new techniques in numerical analysis, in computer arithmetic, and in computer systems. We refer to such a computational environment as one that furnishes *self-validating numerics.*

In this monograph we combine three recently developed approaches in computation to provide a methodology for *self-validating numerics for function space problems* (e.g., differential equations, integral equations, functional equations).

These three approaches are

    (i)   *E-methods*:
            the use of fixed-point theorems to furnish existence, uniqueness, and bounds of good quality for the solution of computational problems [2], [5], [9], [10], [14].

    (ii)   *Ultra-arithmetic*:
            the use of series expansion techniques as an arithmetic methodology [7] and

    (iii) *Computer arithmetic*:
            a precise formulation of floating-point arithmetic for contemporary scientific data types [2], [11].

The term *validation* or *validation process* will be used informally to describe computations performed in this setting and with this methodology. Correspondingly, the term *self-validating methods* will be used to replace the term *E-methods*. Occasional use of the latter term is made for connection to the literature [9]. The quality of the validation process is refined by methods of residual correction. Thus computation of high accuracy and guarantees of the accuracy for such problems are supplied by the computer. Hence in many cases even qualitative analytic information, such as

> containment (bounds)
>
> existence and
>
> uniqueness

of the solution of the exact function space problem being dealt with computationally is supplied by the computer, an additional feature of self-validating numerics. We stress that this type of self-validating information is supplied by computation for the original function space problem itself and not necessarily for some discretized version of the original (although it might very well accomplish the latter as well).

E-METHODS

Proceeding solely from the problem data itself, an E-method furnishes *existence, uniqueness,* and *bounds for the solution* that is sought. This method does not replace conventional numerical methods. Rather the conventional methods are used to supply a residual correction iteration framework and a starting approximation. The E-method uses these with appropriate rounding and arithmetic techniques to simulate a fixed-point iteration (cf. [10]). When the iteration halts, the favorable validation properties are produced. We discuss at length the fixed-point theorems and associated computational point of view.

In the function-space context, the E-method produces a set of functions inside of which the solution being sought is guaranteed to exist. The set is described by its boundaries, the latter being composed of computer-representable functions. The diameter of the set of functions produced is typically small, and it is this property that assures high quality of the bounds. Residual correction methods in the function-space context and the arithmetic framework that we deal with are based on an analogy between floating-point numbers with expansion in terms of basis functions.

The problem may be unsolvable or so poorly conditioned that the iteration fails or is stopped by a resource limitation. Even in this case the computer user is provided with firm information about his problem. The conventional situation in which the computer user is supplied with an approximation and no statement of its quality may be avoided.

To be sure, E-methods have greater cost, but experience has shown that this cost is typically not large compared to the cost of the conventional part of the computation. The costs increase as the problem condition degrades. These added costs are to be weighed against the cost of proceeding without validation. Or they may be compared with having the computer itself provide the validation, etc., as opposed to conventional a priori estimates of quality. We recall that a priori estimates in numerical analysis are typically as difficult to provide as the problem solution itself, and even when they are available, they typically furnish bounds that are so pessimistic as to be of little practical use. Indeed the dependence of the computer user on traditional a priori concepts of the numerical analysis of function-space problems, such as local and global truncation errors, is largely eliminated since with the computed result now come a posteriori statements of existence and quality. Last, the human and computer cost of reruns with parameter changes (itself no certain technique) is eliminated.

ULTRA-ARITHMETIC

Contemporary scientific computation employs a variety of data types such as reals (actually floating-point numbers in a given range), vectors, matrices, and complex versions of these as well as intervals over all of these. The methods of numerical analysis have generated many of these data structures and types as well as processing requirements associated with them. Numerical analysis itself finds its procedures in turn evolving from the body of mathematical methodology, and in a sense, is the bridge between that methodology and scientific computation. Taken in this way mathematical methodology amounts to a limited set of operations, consisting more or less of numerical algebra applied to the limited collection of data types just enumerated. This limitation of approach unnecessarily curtails the applicability of both mathematics and of digital computation. The operations and constructs of mathematics that can be implemented directly in digital computers are far greater in number than those that are currently implemented. The digital computer can be made to appear as a more accurate image of mathematical constructs and operations than it now is. This is the viewpoint of ultra-arithmetic, where the structures, data types, and operations corresponding to functions are developed for direct digital implementation.

The passage from arithmetic to ultra-arithmetic is made conceptually simple by means of an analogy between floating-point numbers and generalized Fourier series, the coefficients of the latter being the analogues of the digits of the former. A digital computer equipped with ultra-arithmetic will be a highly congenial tool for computation. Problems associated with functions will be solvable on computers just as we now solve algebraic problems.

COMPUTER ARITHMETIC

A precise definition of computer arithmetic is of interest in its own right, but it forms as well a foundation on which ultra-arithmetic and E-methods may be developed. In the theory of computer arithmetic [11], arithmetic operations are introduced for computer data types by a mapping principle called semimorphism. This principle employs roundings to define computer arithmetic in terms of actual (i.e., full-precision) arithmetic. In the same way we introduce a collection of spaces that contain the various function data types of ultra-arithmetic. We also introduce appropriate rounding or projection operators that map these spaces into computer-realizable manifolds. The semimorphism mapping principle is then used in conjunction with these roundings to define the algebraic operations of ultra-arithmetic (i.e., computer versions of arithmetic for functions and of integration).

Indeed, if $\mathcal{M}$ denotes a function space and $S_N: \mathcal{M} \to S_N(\mathcal{M})$ a projection or rounding operator of $\mathcal{M}$ onto a finite-dimensional manifold $S_N(\mathcal{M})$ of $\mathcal{M}$, the algebraic structure $(\mathcal{M}; +, -, \bullet, /, \int)$ of $\mathcal{M}$ is mapped into the computer structure $(S_N(\mathcal{M}); \boxplus, \boxminus, \boxdot, \boxslash, \boxintegral)$ by means of the semimorphism principle. This new structure is called a *functoid*.

Among the function data types are interval types, and among the roundings are directed roundings. These constructs supply a precise interval ultra-arithmetic. The E-methods in turn employ this arithmetic to establish the mapping of a set of functions into itself, which is required by the fixed-point theory. It is essential that this precise interval ultra-arithmetic and the directed roundings be supplied. Indeed, self-validating methods furnish mathematical properties with guarantees. It is impossible to supply these precise conclusions without a precise statement and a correspondingly appropriate implementation of the arithmetic.

What is the relation between the methodologies introduced in these pages and the large body of numerical methods that already exist for function-space problems, such as Galerkin methods, spectral methods, the tau-method, and the like ([6], [8], [13]). The existing methodologies are not displaced, but augmented by our point of view. The existing methods are theoretical, typically with a priori analyses and are conceived in a framework of an "infinite-precision" computer. A large part of what we do is to use these function-space methods with an arithmetic point of view. That point of view, comprising the three aspects just discussed, treats the computation of function-space problems directly in a "finite-dimensional" computer[†] with a posteriori specification of the results. The traditional methods may still be used, but the full benefits of high precision and guarantees are not supplied by their use unless the modifications to be discussed here are also exercised. On the other hand, our "finite-dimensional" computer supplies digital analogues of function-space constructs, and so, even a traditional approach to numerical analysis acquires augmented capability.

Making our methods readily accessible to computer users requires, of course, precisely defined arithmetic and algebraic operations along with the appropriate roundings. A higher programming language equipped with all of these constructs will then make the use and exploitation of our methods congenial to the user.

In Chapter 2 we introduce mathematical preliminaries that consist principally of fixed-point theorems and the computational context in

---

[†] A bona fide computer, in fact, when supplied with a third of these three aspects.

which we cast them for the development of validating methods in function spaces.

In Chapter 3 we develop the point of view of ultra-arithmetic and introduce the constructs of function-space arithmetic: spaces, bases, roundings, and approximate operations. This is done both for the pointwise case and for the processes of inclusion (directed roundings, etc.).

In Chapter 4 we apply the constructs of Chapter 3 to a collection of sample functional equations and develop thereby a collection of computational tools for application. We treat both linear and nonlinear problems. Once again, the pointwise approach is given, followed by methods for determining inclusions.

In Chapter 5 we develop iterative residual correction (IRC) methods for functional problems. IRC is the tool for refinement of an inclusion so that guarantees, i.e., inclusions of high accuracy become possible.

In Chapter 6 comments are given on the requirements of a programming language that are needed to make the tools and constructs of our methodology conveniently available in actual practice on a computer.

In Chapter 7 the techniques for adapting our methodology to a computer are discussed in detail. Then self-validating computational results for specific problems are described.

A pair of glossaries and a list of references conclude the monograph. The first is a glossary of symbols used while the second is a glossary of concepts.

REMARK ON ENUMERATION

The chapters, sections, and subsections are enumerated in the conventional lexicographical manner, as may be seen from the table of contents.

The numbering of equations, which restarts for every change in chapter, section, or subsection, is also lexicographical. An equation is given the chapter, section, or subsection number, then a hyphen, and then its number within that chapter, section, or subsection. Thus, for example, (4.1.1-10) refers to the tenth enumerated equation in Subsection 4.1.1; (4.1-3) refers to the third enumerated equation in the part of Section 4.1 that precedes Subsection 4.1.1; (4-2) refers to the second enumerated equation in the part of Chapter 4 that precedes Section 4.1; etc.

SUGGESTIONS TO THE READER

The level of the material in this monograph varies greatly. There are many involved concepts and many illustrations and examples. As an aid to the reader in gaining an overview, we propose a subset of the monograph for a first reading:

Chapter 1: all.

Chapter 2: through Section 2.1 and Theorem 4 of Section 2.2.1.

Chapter 3: through Section 3.1.2(i) Chebyshev rounding and
               Section 3.2 up to directed Chebyshev rounding.

Chapter 7: through Section 7.2.2

We further suggest that certain parts of the monograph be read last: Section 2.2.2, Section 3.1.2(ii) and Section 3.2.2(ii).

Chapter 2

MATHEMATICAL PRELIMINARIES

Fixed point theorems lie at the basis of self-validating methods (E-methods (cf. [10])). These are computational methods that specify an interval, that is, an upper and a lower bound in which we may assert that the solution is to be found. Since we are concerned with problems whose solutions are functions (e.g., differential and integral equations), the computational setting is a function space $\mathcal{M}$, in particular, a Banach space. As we shall see, the computation itself will occur in some finite-dimensional subspace $S(\mathcal{M})$ of $\mathcal{M}$ and the bounds to be produced are themselves functions in that manifold.

We begin in Section 2.1 with a basic formulation of a self-validating method in $\mathcal{M}$. The Schauder-Tychonoff fixed point theorem is stated, and then the computational context is introduced and explained. In Section 2.2 we deal with a broader setting for self-validating methods. We discuss a number of notions and theorems that provide the basis for more general implementation of self-validating methods. The theorems include modifiations of traditional fixed point theorems appropriate to our purpose. Some of these ideas are well known, but we include them for the sake of unity in our exposition and for the convenience of the reader.

2.1 BASIC FORMULATION OF SELF-VALIDATING METHODS IN $\mathcal{M}$

To begin, we recall the following fixed point theorem,[†] dealing with elements of $\mathbf{P}\mathcal{M}$, the power set of $\mathcal{M}$.

THEOREM (Schauder–Tychonoff)

Let $\mathcal{M}$ be a Banach space and $F\colon \mathcal{M} \to \mathcal{M}$ a compact mapping. Let $U \in \mathbf{P}\mathcal{M}$ be a nonempty, convex, closed, and bounded set such that

$$FU \subset U. \qquad (2.1\text{-}1)$$

Then there exists a fixed point of F in FU, a fortiori in U.

COMPUTATIONAL CONTEXT FOR FIXED POINT THEOREMS

An iteration scheme is the standard method for numerical implementation of a fixed point theorem. So we take the computation to be the recurrence

$$Y_{i+1} = F(Y_i), \quad i = 0,1,\dots \qquad (2.1\text{-}2)$$

with Y_0 given in $\mathbf{P}\mathcal{M}$. $F\colon \mathbf{P}\mathcal{M} \to \mathbf{P}\mathcal{M}$ so that all $Y_i \in \mathbf{P}\mathcal{M}$, $i = 0,1,\dots$. We suppose that these sets Y_i are nonempty, convex, closed, and bounded. That is, the Y_i satisfy all of the hypotheses of the fixed point theorem except for (2.1-1), the intoness. This state of affairs must be arranged a priori and is dependent on F and the definition of $\mathcal{M}$.

If during the course of the computation (2.1-2) $Y_{i+1} \subset Y_i$ occurs, the remaining hypothesis of the fixed point theorem is established. The

---

[†] For convenience we shall always formulate the fixed point theorems in this paper in the setting of a Banach space. However, we note here that the more general setting of a locally convex topological vector space is appropriate as well for these theorems.

iteration is considered to have succeeded and is (a posteriori) appropriate for the computational purpose. In particular, the set Y_i produced computationally contains the solution we seek. Thus, the computation provides the existence as well as a bound for the solution. Uniqueness would also be provided if in addition to $Y_{i+1} \subseteq Y_i$ an additional property such as contraction were to be computationally established. In the event of such a contraction, a theorem such as the Picard fixed point theorem would enable us to assert the uniqueness of the solution in Y_{i+1}. The contraction property is not mandatory. Indeed, we shall give theorems that furnish uniqueness a posteriori and that do not employ contraction. We shall return to this point in Section 2.2.

All this information about the solution is provided computationally and so may be viewed as a benefit that defrays the cost of the computation (2.1-2), which is certainly more costly than the standard computational approach.

Of course there remains to provide the means to choose F and $\mathcal{M}$. Choices for F and $\mathcal{M}$ will be discussed later, but we shall comment on possibilities for F here. The criteria required of F are to allow reasonable likelihood for the intoness property to hold (or the contractivity, as the case may be).

We distinguish two cases for the choice of F (the explicit and the implicit), which we now describe.

The Explicit Case

The problem to be solved is taken in the form

$$y = f(y) \tag{2.1-3}$$

with the operator f a good candidate to be contracting. We take the iteration

$$y_{i+1} = f(y_i), \ i \geq 0, \tag{2.1-4}$$

with y_0 given in $\mathcal{M}$.

If $f \in C^1$ we transform (2.1-4) into the following form, which will be required by the fixed point theorems in Section 2.2:

$$Y_{i+1} := f(\tilde{y}) + f'(Y_i \,\underline{\cup}\, \tilde{y})(Y_i - \tilde{y}), \qquad (2.1\text{-}5)$$

with $Y_0 = \tilde{y}$.

Here $\tilde{y}$ is a given approximation to the solution of (2.1-3), perhaps obtained from (2.1-4). (2.1-5) is an iteration in $\mathbf{P}\mathcal{M}$, the power set of $\mathcal{M}$, and we perform this iteration until an index k is reached so that $Y_{k+1} \subseteq Y_k$. We shall discuss the computational implementation of (2.1-4) and (2.1-5) in Chapter 7. If such an index k is not reached ("with a reasonable amount of computing"), we seek to replace (2.1-5) by some higher-order method. Such an approach is contained in the following description of the implicit case.

The Implicit Case

The problem to be solved is taket in the form

$$g(y) = 0. \qquad (2.1\text{-}6)$$

This is transformed into

$$y = f(y) := y - \mathcal{Q}g(y), \qquad (2.1\text{-}7)$$

where $\mathcal{Q}$ is an appropriate linear operation. For example,

$$\mathcal{Q} \sim (g'(y))^{-1} \quad \text{(Newton-type method)},$$
$$\mathcal{Q} \sim (u - v)(g(u) - g(v))^{-1} \quad \text{(a secant-type method)}.$$

(2.1-7) is now an explicit case.

2.2 A BROADER SETTING FOR SELF-VALIDATING METHODS

The direct implementation of self-validating methods described in Section 2.1 limits the scope of such methods. In the context of fixed point methodology one or another hypothesis on the mapping F and sets being mapped is frequently difficult to establish. For this purpose, a number of specialized variations of fixed point theorems have been invented.

On the other hand, more explicit information about the mapping such as properties of its resolvent or spectrum may be exploited for purposes of establishing a fixed point. In this section we formulate a number of such assertions. Of course, our framework will be oriented toward the computational setting that we have in mind.

We begin by introducing some notation and certain concepts and then we continue with the fixed point theorems. The latter are divided into categories: so called basic fixed point theorems treated in Section 2.2.1 and then a modified Krasnoselski-Darbo fixed point theorem treated in Section 2.2.2.

We shall use the symbol $\overset{\circ}{\subset}$ to denote *strict inclusion* in $\mathcal{M}$. That is, given M_1 and $M_2 \subset \mathcal{M}$, then

$$M_1 \overset{\circ}{\subset} M_1 :\Longleftrightarrow \overline{M}_1 \subset \overset{\circ}{M}_2, \tag{2.2-1}$$

i.e., the closure of M_1 lies in the interior $\overset{\circ}{M}_2$ of M_2.

ℓ denotes a linear mapping of $\mathcal{M}$ into $\mathcal{M}$ and $\mathcal{L}$ a (bounded and convex) set of linear operators each of which takes $\mathcal{M}$ into $\mathcal{M}$. E will denote the identity mapping. $\lambda(\ell)$ denotes an eigenvalue of ℓ, and $\Lambda(\ell)$ denotes the set of all such eigenvalues. The set $|\Lambda(\ell)|$ is defined as follows:

$$|\Lambda(\ell)| := \{|\lambda(\ell)| \mid \lambda(\ell) \in \Lambda(\ell)\}. \tag{2.2-2}$$

We now formalize the notion of inclusion of a function in a *tube of functions,* i.e., inclusion in a pointwise sense.

Let $\mathcal{M}$ be the set of bounded real-valued functions of a real variable. For $f \in \mathcal{M}$, let $U_x(f(x))$ be an open neighborhood of $f(x)$ (the value of f at x). Then

$$U(f) = \{g \in \mathcal{M} \mid \bigwedge_x g(x) \in U_x(f(x))\} \tag{2.2-3}$$

is a set of functions (a tube) containing $f(x)$, pointwise.

We recall two well-known concepts associated with a subset U of a Banach space that we shall make use of. First, we say that U is balanced if for any complex number σ, with $|\sigma| < 1$, we have $\sigma U \subset U$. Second, for y an element of the Banach space

$$\|y\|_U := \inf\{t > 0 \mid y \in tU\}$$

is called the Minkowski functional (indexed by U). Recall also that the Minkowski functional is a norm in the case that U is balanced.

2.2.1 Basic Fixed Point Theorems

The modifications of the Banach fixed point theorem and of the Schauder-Tychonoff fixed point theorem that we shall introduce here require two preliminary technical results, Theorems 1 and 2, which we now turn to.

THEOREM 1

Let $\mathcal{M}$ be a Banach space and $\ell : \mathcal{M} \to \mathcal{M}$ a linear mapping. Let $\{0\} \overset{\circ}{\subset} U$, where U is a closed and bounded subset of $\mathcal{M}$. Suppose that

$$\ell U \overset{\circ}{\subset} U. \qquad\qquad (2.2.1\text{-}0)$$

Then

(a) The resolvent $(E - \ell)^{-1}$ exists, at least on U, and both ℓ and $(E - \ell)^{-1}$ are continuous.

(b) Either $\Lambda(\ell) = \phi$ or $|\Lambda(\ell)| < 1$.

(c) If U is balanced, then $\|\cdot\|_U$ is a norm and $\|\ell\|_U < 1$. Furthermore, for all $n \in \mathbf{N}$ with

$$\ell^{n+1}(U) \overset{\circ}{\subset} \ell^n(U),$$

we have $\|\ell\|_{\ell^n(U)} < 1$.

(d) If $(2.2.1\text{-}0)$ is replaced by $\ell U \overset{\circ}{\subset} \kappa U$ for some $\kappa \in \mathbb{C}$, then the conclusions (a)-(c) of this theorem remain valid by replacing ℓ by ℓ/κ.

Proof:

(a) To demonstrate the existence of $(E - \ell)^{-1}$ on U, suppose that p is a nonnull element of U such that $(E - \ell)p = 0$. Since $\{0\} \overset{\circ}{\subset} U$, $q = \sigma p \in U$ for all sufficiently small scalars σ. Since U is closed and bounded, there exists a scalar $\bar{\sigma}$ such that $\bar{q} = \bar{\sigma}p \in \partial U$. Then

$$\ell\bar{q} = (E - (E - \ell))\bar{q} = \bar{q} \in \partial U.$$

This is a contradiction since the left member here is in the interior of U. Thus $E - \ell$ annihilates no nontrivial element of U, and so $(E - \ell)^{-1}$ exists on U.

Since ℓ is bounded on U, it is continuous there. Similarly $(E - \ell)^{-1}$ is continuous.

(b) If $\ell \not\equiv 0$, then $\Lambda(\ell) = \{0\}$. Suppose $\ell \neq 0$ and that $\Lambda(\ell) \neq \phi$. Then let $\lambda \in \Lambda(\ell) \subset \mathcal{C}$ be an eigenvalue (which we may suppose is nonnull) and let v be a corresponding eigenvector so normalized that $v \in U$. Consider the set $\Psi := \{\psi \in \mathcal{C} \mid \psi v \in U\}$. Ψ is closed since U is closed. Thus $|\psi^*| := \max_{\psi \in \Psi} |\psi|$ exists and moreover $\psi^* \in \Psi$. Thus $\psi^* v \in U$ and so by hypothesis $\ell(\psi^* v) \overset{\circ}{\subset} U$, i.e.

$$\ell(\psi^* v) = \psi^* \ell v \overset{\circ}{\subset} U.$$

Then there exists a $\sigma \in \mathbb{R}$ with $\sigma > 1$ so that

$$\sigma\psi^*\lambda c \in \partial U \subset U.$$

Then by the definition of the set Ψ, we note that $\tilde{\psi} := \sigma\psi^*\lambda \in \bar{\Psi}$. But then

$$|\psi^*| \geq |\tilde{\psi}| = |\sigma\psi^*\lambda| = \sigma|\psi^*||\lambda|.$$

Thus

$$|\lambda| \leq \frac{1}{\sigma} < 1.$$

Since λ is arbitrary, $|\Lambda(\ell)| \leq q < 1$ with $q = 1/\sigma$.

(c) Since U is balanced, the Minkowski functional $\|y\|_U$ is a norm. Moreover, $\|v\|_U \leq 1$ for $v \in U$ but not otherwise. For an arbitrary $v \in U$, $\{\ell v\} \overset{\circ}{\subset} U$. Thus there exists $t, q \in \mathbb{R}$ with $t \leq q < 1$ such that $\ell v \in tU$ and $\| \ell v \|_U \leq q < 1$. Combining we have

$$\| \ell \|_U = \sup_{\| v \|_U \leq 1} \| \ell v \|_U = \sup_{v \in U} \| \ell v \|_U < 1,$$

the first assertion of (c).

Clearly for arbitrary continuous linear ℓ, we have $\{0\} \overset{\circ}{\subset} \ell^n(U)$, and moreover, that $\ell^n(U)$ is balanced. Thus the second assertion of (c) follows along the lines of the first.

(d) This is evident since $\ell U \overset{\circ}{\subset} \kappa U$ implies $\ell / \kappa\, U \overset{\circ}{\subset} U$. ■

Remarks:

1. From conclusion (b) of Theorem 1, we may infer that

$$\sigma(\ell) := \sup_{\lambda \in \Lambda(\ell)} | \lambda | = q < 1.$$

Then from the extremal property of $\sigma(\ell)$ with respect to the class of operator norms, there exists a norm (in general unknown) such that

$$\| \ell \| \leq q + \varepsilon < 1$$

for arbitrary and sufficiently small ε. Thus ℓ is a contraction operator (in a norm which is unknown).

2. Without the hypothesis of balance for U, the Minkowski functional is not a norm. However, the property $\| \ell \|_U \leq q < 1$ prevails. Loss of the norm property appears to result in the loss of contractivity. However, uniqueness is achieved by alternative means, as will be seen in the fixed point theorems to follow.

The following Theorem 2 is a preparatory result, which shows that the nonsingularity of the composition $\ell_1 \circ \ell_2$ implies the nonsingularity of ℓ_1 and ℓ_2 separately.

THEOREM 2

Let the Banach space $\mathcal{M}$ and the subset U be as in the hypothesis of Theorem 1. Let ℓ_2 be an arbitrary linear mapping, and let $\mathcal{L}_1$ be the following set of mappings:

$$\mathcal{L}_1 := \{\ell_1 \mid \ell = E - \ell_1 \circ \ell_2 \text{ and } \ell U \overset{\circ}{\subset} U\}.$$

If $y \in U$ is annihilated by each $\ell_1 \in \mathcal{L}_1$, then $y = 0$.

Proof:

Using Theorem 1.a we see that

$$(E - (E - \ell_1 \circ \ell_2))^{-1} = (\ell_1 \circ \ell_2)^{-1}$$

exists on U. Then ℓ_2^{-1} exists on U also. Since

$$(E - \ell_1 \circ \ell_2)U \overset{\circ}{\subset} U$$

then for sufficiently small $\varepsilon > 0$, we have

$$(E - (\ell_1 + \varepsilon E) \circ \ell_2)U \overset{\circ}{\subset} U.$$

Thus, if $u \in (E - \ell_1 \circ \ell_2)U$ and $\varepsilon \ell_2(U) \subset \underset{\sim}{\mathcal{U}}(0)$ (a neighborhood of zero), then $\underset{\sim}{u} + \mathcal{U}(0) \overset{\circ}{\subset} U$. By definition $\ell_1 = \ell_1 + \varepsilon E \in \mathcal{L}_1$. Using $\ell_1 y = 0$ and $\ell_1 y = 0$, we have $\varepsilon y = 0$. Thus, $y = 0$. ∎

We now state and prove a modified version of the Banach fixed point theorem. The modification concerns the form of the hypotheses (ii) and (iii) which specify the contractive nature of the mapping whose fixed point is characterized. The modifications are designed to permit verification of contraction in a computational framework to be dealt with below.

THEOREM 3

Let $\mathcal{M}$ be a Banach space, and let $\mathcal{M} \supset Z \supset Y$ where Y is a closed nonempty subset. Let $f: Z \to \mathcal{M}$ be a mapping, and let $\mathcal{L}(Y)$ be a set of linear operators taking $\mathcal{M}$ into itself. (Y may be regarded as a parametrization of the set $\mathcal{L}$). Let the following additional conditions prevail:

(i) $f(Y) \subset Y$.

(ii) $\bigwedge\limits_{y_1, y_2 \in Y} \bigvee\limits_{\ell \in \mathscr{L}(Y)} f(y_1) - f(y_2) = \ell(y_1 - y_2)$.

(iii) There exists a subset of V of $\mathscr{M}$ with $\{0\} \overset{\circ}{\subset} V$ such that

$$\mathscr{L}(Y)V \overset{\circ}{\subset} V$$

(alternatively there exists an $n \in \mathbf{N}$ such that $\mathscr{L}^{n+1}(Y)V \overset{\circ}{\subset} \mathscr{L}^n(V)$).

(iv) Either (α) V is balanced or (β) f is compact.

Then

(a)   For each $\ell \in \mathscr{L}$, either $\Lambda(\ell) = \phi$ or $|\Lambda(\ell)| \leq q < 1$, and in the case ($\beta$) (i.e., that f is compact) the resolvent $(E - \ell)^{-1}$ exists.

(b)   There exists exactly one fixed point $\hat{y}$ of f and $\hat{y} \in f(Y)$.

In the case that V is balanced we have the following additional conclusions.

(c)   Depending on which two cases in (iii) hold, we have

$$\sup_{\ell \in \mathscr{L}(Y)} \| \ell \|_V = \| \mathscr{L}(Y) \|_V \leq q < 1$$

(alternatively $\| \mathscr{L}(Y) \|_{\mathscr{L}^n(Y)} \leq q < 1$).

(d) F is contracting in the norm $\| \cdot \|_V$.

Proof:

We may suppose without loss of generality that $\mathscr{L}(Y)$ is a closed set so that $\overline{\mathscr{L}}$ fulfills (iii) as well.

(a)   This follows from (iii) and conclusions (a) and (b) of Theorem 1.

(b)   Case (β): The existence of a fixed point $\hat{y}$ follows from (i) and the Schauder fixed point theorem. For the unicity of $\hat{y}$, we argue as follows: Let y_1 and y_2 both be fixed points of f. Then using (ii) we have for some $\ell \in \mathscr{L}$ that

$$y_1 - y_2 = f(y_1) - f(y_2) = \ell(y_1 - y_2),$$

so that

$$(E - \ell)(y_1 - y_2) = 0.$$

Then (a) implies that $y_1 - y_2 = 0$, i.e., the unicity.

Case (α): The existence and uniqueness of a fixed point follows from (i) and (d), which together supply the hypotheses of the Banach fixed point theorem.

Thus, (b) is proved when (d) is established. We turn now to (c) and (d).

(c) For each $\ell \in \mathcal{L}(Y)$, we have according to (iii) that $\ell V \overset{\circ}{\subset} V$. Then from (c) of Theorem 1 we deduce that $\| \ell_V \| \leq q < 1$. Then

$$\|\mathcal{L}(Y)\|_V = \sup_{\ell \in \mathcal{L}} \| \ell \|_V \leq q < 1,$$

where we have used the fact that $\mathcal{L}$ is closed. Thus (c) is established.

(d) Using (ii), we have for each pair $y_1, y_2 \in Y$ an $\ell \in \mathcal{L}$ exists so that $f(y_1) - f(y_2) = \ell(y_1 - y_2)$. From this in turn

$$\|f(y_1) - f(y_2)\|_V = \| \ell(y_1 - y_2) \|_V \leq \| \ell \|_V \| y_1 - y_2 \|_V.$$

Using (iii), we have that $\| \ell \|_V < 1$. Thus (d) follows from the Banach fixed point theorem. ■

We now comment on Theorem 3.

Remarks:

0. In some cases ℓ has the form $\ell = \mathcal{Q}g$ as in (2.1-7). Then the existence of a fixed point y (i.e., $y = f(y)$) implies that $\mathcal{Q}g(y) = 0$. Now use of Theorem 4(b) shows that the operator product $\mathcal{Q}g'$ is nonsingular. Then Theorem 2 implies that both $\mathcal{Q}$ and g' are nonsingular. Hence $\mathcal{Q}g(y) = 0$ implies that $g(y) = 0$.

1. The set of operators $\mathcal{L}(Y)$ may not be representable in practice. Thus for computational purposes, $\mathcal{L}(Y)$ is then to be replaced by a representable superset of operators for which the hypotheses of the theorem are to be satisfied, in particular, the intoness conditions in

(iii) become more stringent.

2. Note that in the case (α) the conclusions (b), (c), and (d) hold in the absence of an estimate for a Lipschitz constant of the mapping f. Moreover, such an estimate need not be known even for the containment of the fixed point since (b) provides this containment.

3. It is in principle possible to compute an a posteriori estimate of the Lipschitz constant from $\|\mathscr{L}(Y)V\|_V$. However, suchlan estimate is typically entirely impractical.

If $j y^{n+1} = f(y^n)$, $n \geq 0$ and κ is the Lipschitz constant for f, then

$$\|\hat{y} - y_0\| \leq \sum_{i=0}^{\infty} \kappa^i \|y_1 - y_0\|.$$

From this we obtain the customary estimate

$$\hat{y} \in y_0 \pm \frac{1}{1-\kappa} \|y_0^1 - y_1\|_V.$$

The estimate is a worst case estimate. By contrast (b) provides estimates for each component.

Theorem 3 requires computation for determining both Y in (i) and V in (iii), and we should, if possible, avoid the need to specify two sets computationally. Indeed, Theorem 3 is relevant in the case (α) since in the case (β) we have a sharper method for computing inclusions. We specify this method in Theorem 4 which is a modification of tte Schauder-Tychonoff fixed point theorem. Theorem 4 delivers the uniqueness of the fixed point. In Theorem 4 we shall make use of $\mathbb{L}(\mathscr{M}, \mathscr{M})$, the set of linear operators taking $\mathscr{M}$ into itself. Corresponding to a set Y we consider a subset $\mathscr{K}(Y) \subset \mathbb{L}$, i.e.,

$$\mathscr{K}(Y) = \{k(y) \in \mathbb{L}(\mathscr{M},\mathscr{M}) \mid y \in Y\}. \tag{2.2.1-1}$$

Thus, Y is a parametrization of the set qf operators $\mathscr{K}$.

THEOREM 4

Let $\mathscr{M}$ be a complex Banach space, and let $\mathscr{M} \supset Z \supset Y$, where Y is a nonempty, convex, closed, and bounded subset of $\mathscr{M}$. Let $f: Z \to \mathscr{M}$

be a compact mapping with the following property. To each $z \in Z$ and to each arbitrary but fixed $g \in \mathcal{M}$ there exists a compact set of compact linear operators $\mathcal{K}(Y) \subset \mathbb{L}(\mathcal{M}, \mathcal{M})$ so that the following conditions prevail:

(i) $\quad \bigwedge_{y \in Y} f(y) \in g + \mathcal{K}(Y)(y - z).$

(ii) $\quad \bigwedge_{y_1, y_2 \in Y} \bigvee_{k \in \mathcal{K}(Y)} f(y_1) - f(y_2) = k(y_1 - y_2).$

(iii) There exists a set $\mathcal{L}(Y)$ of linear operators so that $\mathcal{K}(Y) \subset \mathcal{L}(Y)$ and

$$g + \mathcal{L}(Y)(Y - z) =: F(Y) \overset{\circ}{\subset} Y.$$

Then

(a) There exists exactly one fixed point $\hat{y}$ of f, and moreover, $\hat{y} \in F(Y)$.

(b) For each compact $\ell \in \mathcal{L}(Y)$, either $\Lambda(\ell) = \phi$ or $|\Lambda(\ell)| \leq q < 1$ and the resolvent $(E - \mathcal{K})^{-1}$ exists on $Y - \hat{y}$.

Remark: Note that in the particular case that $z \in Y$, the hypothesis (i) is implied by the hypothesis (ii).

Proof:

(a) existence

From (i) and (iii) we have $f: f(Y) \subset Y$. Indeed, the hypothesis of the theorem then allows us to apply the Schauder fixed point theorem, from which we conclude the existence of a fixed point $\hat{y} \in f(Y)$.

uniqueness

To each compact $\ell \in \mathcal{L}(Y)$ we associate the affine mapping $h(y)$

$$h(y) := g + \ell(y - z): Z \to \mathcal{M}.$$

From (iii)

$$h(Y) = g + \ell(Y - z) \subset F(Y) \overset{\circ}{\subset} Y. \qquad (2.2.1\text{-}2)$$

Thus, by the Schauder fixed point theorem h has a fixed point

$y^* = y^*(\ell) \in Y$, i.e.,

$$h(y^*) = g + \ell(y^* - z) = y^*. \qquad (2.2.1\text{-}3)$$

Let $U := Y - y^*$ so that $\{0\} \overset{o}{\subset} U$. Now subtracting (2.2.1-3) from (2.2.1-2), we get

$$\ell(Y - y^*) = \ell(Y - z) - \ell(y^* - z) \subset Y - y^*,$$

i.e.,

$$\ell U \overset{o}{\subset} U. \qquad (2.2.1\text{-}4)$$

Now from Theorem 1(a) we have that $(E - \ell)^{-1}$ exists. Suppose y_1 and y_2 are fixed points of f. Then from (ii) there exists a mapping $\ell \in \mathscr{K}$ such that

$$y_1 - y_2 = f(y_1) - f(y_2) = \ell(y_1 - y_2),$$

i.e., $(E - \ell)(y_1 - y_2) = 0$. Since $(E - \ell)^{-1}$ exists, $y_1 - y_2 = 0$.

(b) Using (2.2.1-4) and Theorem 1(a) and (b), we have that either $\Lambda(\ell) = \phi$ or $|\Lambda(\ell)| < 1$ and the resolvent exists on $U = Y - y^*$.

■

We now comment on Theorem 4.

Remarks:

0. Remark 0 following theorem 3 is valid here.

1. Referring to Remark 1 following Theorem 1 we see that for each $k \in \mathscr{K}$ there exists a norm $\|\cdot\|_k$ (not explicitly known) with respect to which k is contractive.

2. Typically the set $\mathscr{K}(Y)$ of linear operators is taken to be the Frechet derivative f' of f. For example, take $y \in Y := A + Bx$ and $\mathscr{K}(Y)\cdot = \int (A + Bt)\cdot dt$ where A and B are intervals of $\mathbb{R}$. Apply $\mathscr{K}(Y)$ to $Y - z$. The result $\mathscr{K}(Y)(Y - x) = \int Y(Y - z)dt = \int (A + Bt)((A - a) + (B - b)t)dt$ is always a compact set. However, this compactness property is not conveyed if the set is only represented by means of its boundaries (e.g., interval polynomials). However, the compactness of $\mathscr{K}$ is a theoretical requirement. Indeed, in practice $\mathscr{K}$ is replaced by a superset $\mathscr{L}(Y)$, which is typi-

cally neither compact nor required to be. In the situation of the example discussed here we have

$$\mathscr{K}(Y)(Y-z) \subset \mathscr{L}(Y)(Y-z)$$

$$:= A(A-a)\int dt + (A(B-b)$$
$$+ B(B-a))\int t\,dt + B(B-b)\int t^2 dt$$

which, while simple in form for computation, is in general not a compact set.

3. The typical application of Theorem 4 is to provide an inclusion of the solution of explicit differential equations and integral equations of the second kind. As a by-product, a proof of existence and uniqueness is also established. For implicit problems, generalizations of Theorem 4 are required, and these are supplied by Theorem 5 and 6 below.

4. Referring to Theorems 6 and 7 we see that compactness in Theorem 4 may be replaced by α-condensing.

2.2.2 Modified Krasnoselski–Darbo Fixed Point Theorems

In the implicit case the property of contraction is not typically available. Rather the function in question is a semicontraction composed of a contracting part and of a compact part. The theorems in this subsection generalize Theorems 3 and 4 for applicability to this situation. We start with the notions of the measure of noncompactness and of an α-condensing operator, which are needed for these theorems.

Let $\alpha(M)$ denote the *measure of noncompactness* of a subset M of a Banach space $\mathscr{M}$, where

$$\alpha(M) := \inf\{\varepsilon > 0 \,|\, M \text{ has a finite covering each set}$$
$$\text{of which is of diameter less than } \varepsilon\}$$

(cf. [15], [17]). An operator $K: M \to M$ is called *α-condensing* if for any bounded subset Ω of M with $\alpha(\Omega) > 0$, we have

$$\alpha(K(\Omega)) < \alpha(\Omega).$$

We now give the fixed point theorem of Krasnoselski-Darbo (cf. [15]).

THEOREM 5

In a Banach space $\mathcal{M}$ a continuous and α-condensing mapping of a closed, bounded, and convex set into itself has at least one fixed point.

It is of interest to review a sufficient condition for an operator to be α-condensing. We deal with a case that is typical in application.

Let A and B be subsets of $\mathcal{M}$. Let $K: A \to B$ be a continuous operator that has the pointwise representation

$$K(x) \equiv V(x,x) \text{ for all } x \in A.$$

Here $V: A \times A \to B$ is continuous,

$$V(\bullet,x): A \blacksquare B$$

is a (partially) compact operator for each fixed $x \in A$, and

$$V(x,\bullet): A \to B$$

is (partially) contracting for each fixed x with a uniform contraction constant $\kappa < 1$. Such operators K are called *strictly semicontractive* in A, and are therefore α-condensing with $\kappa < 1$ (cf. [15]).

As an example, we note that the sum of a compact operator and of a contracting operator is an α-condensing operator. For instance, the operator

$$y(t) = g(y) + \int_0^t f(\tau, y(\tau)) d\tau$$

$(V(\bullet,y) = g(y) + \int_0^t f(\tau,\bullet) d\tau$ and $V(y,\bullet) = g(\bullet) + \int_0^t f(\tau, y(\tau)) d\tau)$ is strictly semicontractive if g has a Lipschitz constant $\kappa < 1$ with respect to its second argument. For further examples, see [15], [17].

We now proceed to Theorems 6 and 7, each of which is a variant of Theorems 3 and 4 appropriate to the α-condensing framework. In particular, Theorem 6 deals with the balanced-set hypotheses (Theorem 3(iii)α), while Theorem 7 deals with the compact mapping hypotheses (Theorem 3(iv)β).

THEOREM 6

(o) Let $\mathcal{M}$ be complex Banach space, and let U and Y be nonempty subsets of $\mathcal{M}$ each of which is convex, closed, bounded, and balanced. In particular, $\{0\} \overset{\circ}{\subset} U$. For arbitrary but fixed $z_1 \in \mathcal{M}$, let $\mathcal{L}_1(y)$ be a set of linear mappings.

Suppose that the following hypotheses are satisfied:

(i)   for all $y \in Y$, $f(y) = h(y,y)$ where

(ii) for all $u \in Y$, $h(u,v)$ is a compact mapping with respect to its second argument v, for all $v \in Y$

(iii) for all $u,v \in Y$, $h(u,v) \in H(u,y) := g(v) + \mathcal{L}_1(Y)(u - z_1)$, for some $g(v) \in \mathcal{M}$

(iv) $H(Y,Y) \subseteq Y$

(v) $\mathcal{L}_1(Y)U \overset{\circ}{\subset} U.$

Then

(a) f is strictly semicontracting on Y, and therefore

(b) there exists a fixed point $\hat{y} \in H(Y,Y)$.

Proof:

Using (v) we obtain $\|\mathcal{L}_1\|_U < 1$ (as in Theorem 1). Combining this with (iii) we deduce that every $h(u,v) = g + \ell(u - z_1)$ with $v \in Y$ and $\ell \in \mathcal{L}_1$ is (partially) contracting with respect to $u \in Y$. Combining this in turn with (ii) we deduce conclusion (a) of the theorem, namely, $f(y) = h(y,y)$ is strictly semicontracting, i.e., $f(y)$ is an α-condensing mapping. Finally, with (iv) all of the hypotheses of Theorem 5 are fulfilled, and so the existence of $\hat{y}$, the fixed point of conclusion (b) follows.   ∎

Some comments concerning this theorem are now given.

Remarks:

1. Theorem 6 does not guarantee uniqueness of the fixed point in Y. Additional conditions for uniqueness are specified in Theorem 7.

2. If the contractivity of $h(u,\bullet)$ with respect to a is known a priori (viz., $\|h(r,\bullet) - h(s,\bullet)\| \le \lambda \|r - s\|$ with $\lambda < 1$), then the hypothesis $f(Y) \subset Y$ (in place of (iv)) suffices for existence of $\hat{y}$.

The following theorem is a specialization of Theorem 6. It combines Theorems 3 and 4 in a general way, and it provides for uniqueness of the fixed point. Theorem 7 is of interest in the case that it is not known a priori that f is an α-condensing mapping, for otherwise Theorem 7 is simply Theorem 4 with "compactness" in the latter replaced by "α-condensing".

THEOREM 7

Let hypotheses (o) (concerning $\mathcal{M}$, U, Y, z_1, g_1, and $\mathcal{L}_1(y)$) of Theorem 6 be valid. Furthermore, for arbitrary $z, z_1, k \in \mathcal{M}$ and for an appropriate function $Q: Y \to \mathcal{M}$, let there exist sets of linear mappings $\mathcal{L}_1(Y)$ and $\mathcal{L}(Y)$ so that the following hypotheses are satisfied:

(i) $\underset{y \in Y}{\bigwedge} f(y) = h(y,y)$.

(ii) $\underset{x,y \in Y}{\bigwedge} \ \underset{\ell \in \mathcal{L}(Y)}{\bigvee} f(x) - f(y) = \ell(x - y)$,

$\underset{y \in Y}{\bigwedge} f(y) \in k + \mathcal{L}(Y)(y - z)$.

(iii) $\underset{u \in Y}{\bigwedge} h(u,v)$ is a compact mapping with respect to its second argument $v \in Y$.

(iv) $\underset{u,v \in Y}{\bigwedge} h(u,v) \in g(v) + \mathcal{L}_1(Y)(u - z_1)$.

(v) $g(Y) + \mathcal{L}_1(Y)(Y - z_1) =: H(Y,Y) \overset{\circ}{\subset} Y$.

(vi) $\mathscr{L}_1(Y)U \overset{\circ}{\subset} U.$

(vii) $k + \mathscr{L}(Y)(Y - z) =: F(Y) \overset{\circ}{\subset} Y.$

Then there exists one and only one fixed point $\hat{y} \in H(Y,Y)$.

Remark: In the case that $z \in Y$, the second condition in (ii) follows from the first.

Proof:

From (iv) and (vi) we deduce that $h(u,v)$ is (partially) contracting with respect to the first argument of h. Combining this observation with (iii), we deduce that f is strictly semicontracting, and thus, f is an α-condensing mapping as well. Then the conclusion follows from Theorem 4 upon substituting α-condensing for compactness.   ■

Chapter 3

ULTRA-ARITHMETIC AND ROUNDINGS

Contemporary scientific computation employs a variety of data types such as reals (actually floating-point numbers), vectors, matrices, and complex versions of these as well as intervals over all of these. The methods of numerical analysis have generated many of these data structures and types as well as processing requirements associated with them. Numerical analysis itself finds its procedures in turn evolving from the body of mathematical methodology, and in a sense, is the bridge between that methodology and scientific computation. Taken in this way mathematical methodology amounts to a limited set of operations, consisting more or less of numerical algebra applied to the limited collection of data types enumerated above. This limitation of approach unnecessarily curtails the power of both mathematics and of digital computation. The operations and constructs of mathematics that can be implemented directly in digital computers are far greater in number than those that are currently implemented. The digital computer can be made to appear as a more accurate image of mathematical constructs and operations than it now is. As an indication that this is indeed the case we display Table 3-1, contrasting features of representing numbers in $\mathbb{R}$ as a decimal expansion and functions in L^2 as a generalized Fourier series. From these in turn, as the table shows, data types for numbers or functions are appropriate roundings from $\mathbb{R}$ or L^2, say, i.e., mappings into a digital computer. This is the viewpoint of ultra-arithmetic, where the structures, data types, and operations corresponding to functions are developed for direct digital implementation. A digital computer equipped with ultra-arithmetic will be a highly congenial tool for computation. Problems associated with functions will be solvable on computers just as we now solve algebraic problems. Moreover, the considerably enlarged set of structures, data types, and operations should make for the generation of far reaching concepts of computer architecture.

In Section 3.1 we introduce the constructs of our methodology: the spaces, bases, roundings, and approximate operations. Examples of each of these are then described in some detail. Then in Section 3.2 we consider the analogous constructs that are to be employed when computation that supplies inclusion is in question. Examples are given in these cases also.

3.1 SPACES, BASES, ROUNDINGS, AND APPROXIMATE OPERATIONS

We will deal with a number of familiar spaces.

$\mathbb{R}^d$ denotes the d-dimensional Euclidean space, and $X \subset \mathbb{R}^d$ denotes a d-dimensional subset.

$\mathbb{R}^k[x]$ denotes the space of all polynomials in x of degree not greater than k. $\mathbb{R}[x] := \bigcup_{k \geq 0} \mathbb{R}^k[x]$.

We also consider the spaces $\mathbb{R}^k[x](X)$, $C^\infty(X)$, $L^2(X)$ and $H^k(X)$, the last being the kth Sobolev space with the norm

$$\|f\|_k = \left(\sum_{|\alpha| \leq k} \int_X |\partial_\alpha f|^2 \right)^{1/2}.$$

Numbers $a \in \mathbb{R}$	**Functions $f \in L^2$**
Representation as a decimal expansion	Representation as a generalized Fourier series
$$a = \sum_{i=M}^{\infty} \frac{a_i}{10^i}$$	$$f = \sum_{j=-\infty}^{\infty} a_j \phi_j$$
"basis" for the expansion $$e_i = 10^{-i}, \; i = M, M+1, \ldots$$	basis for the expansion $$\phi_j = e^{ijx}, \; j = 0, \pm 1, \ldots$$
"coefficients" $$a_i \in \{0, 1, \ldots, 9\}$$	coefficients $$a_j = \langle f, \phi_j \rangle \, / \sqrt{2\pi}$$ $$= \frac{1}{\sqrt{2\pi}} \int_{-\pi}^{\pi} f\phi_j \, dx$$
"normalized basis" and "coefficients" $$\tilde{e}_i = 1$$ $$\tilde{a}_i \in \{0, 10^{-i}, \ldots, 9 \times 10^{-i}\}$$	normalized basis and coefficients $$\tilde{\phi}_j = \frac{1}{\sqrt{2\pi}} \phi_j \quad \tilde{a}_j = a_j$$
The $\tilde{a}_i$. decay exponentially with i.	The $\tilde{a}_j$ decay like the pth power provided f is periodic and p-times differentiable.
Rounded number $$S_N a := \sum_{i=M}^{N} \tilde{a}_i \tilde{e}_i$$ S_N is some rounding operator.	Rounded function $$S_N f := \sum_{j=-N}^{N} a_j \phi_j$$ The rounding operator S_N corresponds to truncation.
data type $$a_M a_{M-1} \cdots a_0 \cdot a_{-1} a_{-2} \cdots a_{-N},$$ a finite string of integers.	data type $$(a_{-N}, a_{-N+1}, \ldots, a_0, \ldots, a_N),$$ a finite string of reals.
Rounding error estimate $$\|a - S_N a\| \leq .5 \times 10^{-N}$$ This estimate follows from the exponential decay of the $\tilde{a}_i$.	Rounding error estimate This estimate follows from the power-like decay of the a_j.

Table 3-1: Contrasting number and function data types

X denotes the domain over which the functions in each particular space is defined. For convenience, we denote the L^2-norm of f by $\|f\|$ instead of $\|f\|_0$. We shall also drop the symbol X when no confusion will result.

Let $\mathcal{M}$ denote a separable Hilbert space. Let $\Phi := \{\phi\}_{\ell=0}^{\infty}$ denote a basis, usually orthonormal (in $\mathcal{M} = L^2(X)$, for example). We use $S_N(\mathcal{M})$ to denote the subspace of $\mathcal{M}$ spanned by $\Phi_N := \{\phi_i\}_{i=0}^{N}$. $S_N(\mathcal{M})$ is called a *screen* of $\mathcal{M}$. S_N denotes an operator called a *rounding*, $S_N: \mathcal{M} \rightarrow S_N(\mathcal{M})$. For each $f \in \mathcal{M}$, we call $S_N f$ the round of f (onto $S_N(\mathcal{M})$), so that $(I - S_N)f$ is the *rounding error*. $(I - S_N)$ is the rounding error operator. (The case of a Fourier series is an exception to this convention, since in that case we use S_N to denote the projection onto the subspace spanned by $\Phi_N := \{e^{inx}\}_{n=-N}^{N}$).

It is important that roundings fulfill some natural conditions. This is demonstrated in [11] for roundings in finite-dimensional spaces, where such conditions as well as the properties of the algebraic structure induced in the so-called *screen* (i.e., the set of computer representable numbers) are characterized by the descriptive term *semimorphism*, for the roundings. While such natural conditions and other analogues have not yet been developed in function spaces, we shall nevertheless and by way of anticipation use the term *semimorphism* for the roundings to be introduced here.

It is not necessary to confine roundings to the class of projections onto linear subspaces. We have made such a restriction here for reasons of clarity. The more general situation is one in which the rounding S_N is a nonlinear mapping onto a subset $S_N(\mathcal{M})$ characterized by N' degrees of freedom, where

$$N' := N + 1$$

(i.e., S_N is a finite nonlinear operator). In any case the following requirement must be met by any (problem-independent) rounding S_N:

$$\bigwedge_{f \in S_N(\mathcal{M})} S_N f = f \qquad (3.1\text{-}0)$$

(invariance of the rounding on the screen). By contrast, in the problem-dependent case, see (3.1.2-29).

For the purpose of representation in a computer, we introduce the *coefficient space* $R_N(\mathcal{M})$ corresponding to $S_N(\mathcal{M})$. We do this by means of an *isomorphism*

$$i: S_N(\mathcal{M}) \to R_N(\mathcal{M}),$$

where the operator

$$R_N \equiv iS_N.$$

Thus, to each rounding S_N there is associated a corresponding rounding $R_N: \mathcal{M} \to R_N(\mathcal{M})$.

We represent these constructs in Figs. 3.1-1 and 3.1-2.

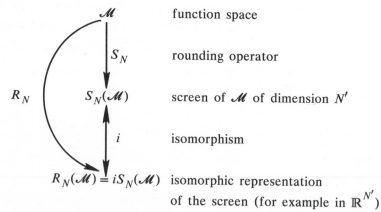

$\mathcal{M}$	function space
S_N	rounding operator
$S_N(\mathcal{M})$	screen of $\mathcal{M}$ of dimension N'
i	isomorphism
$R_N(\mathcal{M}) = iS_N(\mathcal{M})$	isomorphic representation of the screen (for example in $\mathbb{R}^{N'}$)

Figure 3.1-1: Spaces, screens and mappings

We shall often abbreviate $S_N f$ by $S_N f = v$ and correspondingly $R_N f = iv$. If no confusion results, we shall write v for iv as well.

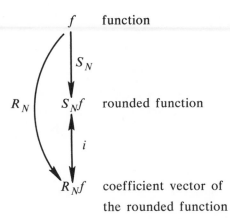

f function

$S_N f$ rounded function

$R_N f$ coefficient vector of
the rounded function

Figure 3.1–2: Functions and rounded functions

An analytic relation between $R_N f$ and $S_N f$ is useful and necessary for computation. We now turn to developing it.

Let $\Phi = \{\phi_0, \phi_1, \ldots\}$ be a basis for $\mathcal{M}$. For each $f \in \mathcal{M}$, we have

$$f = \Phi * a := \sum_{i=0}^{\infty} \phi_i a_i, \qquad (3.1\text{-}1)$$

$a \in k(\mathcal{M})$, the space of sequences (in $\mathbb{R}$ or $\mathbb{C}$) for which the series (3.1-1) is convergent. a is also to be interpreted presently as a formal infinite column vector. Note that ϕ_j corresponds to the sequence $a^j = \{\delta_{ij}\}_{i=0}^{\infty} = R_N \phi_j$, $j \leq N$, i.e., $\phi_j = \Phi * a^j$.

Putting

$$S_N f = S_N(\Phi * a) = S_N(\Phi) * a =: \Phi * \mathcal{A}(S_N) * a,$$

we define an $N' \times \infty$ matrix $\mathcal{A}(S_N)$. $S_N(\phi_j)$ is the jth column of $\mathcal{A}(S_N)$ so that $S_N(\phi_j) = \Phi * R_N(\phi_j)$. Thus, we can write

$$S_N f = S_N(\Phi * a) = \Phi * \mathcal{A}(S_N) * a = \Phi * R_N f, \qquad (3.1\text{-}2)$$

and we may make the following identification: $R_N f = \mathcal{A}(S_N) * a = iv$. The matrix $\mathcal{A}(S_N)$ plays the central role for actual computations. Indeed, the isomorphism between S_N and R_N is represented by and is implementable numerically through the matrix $i(S_N) = \mathscr{A}(S_N)$.

Remark: In applications the vector a in (3.1-2) has only finitely many, say q, leading nonzero components. Then in the product $\mathscr{A}(S_N)*a$, it suffices to take the corresponding q leading columns, neglecting thereby all formally zero terms. We shall hereafter, in fact, do this, but we shall not employ a different notation for $\mathscr{A}(S_N)$. The context should eliminate any confusion.

In every case rounding results in a finite series;

$$S_N f = \Phi * R_N f, \tag{3.1-3}$$

since $R_N f = \mathscr{A}(S_N)*a \in R_N(\mathscr{M})$ is a finite vector. Thus, it usually suffices to consider only the finite basis $\Phi_N = \{\phi_0,...,\phi_N\}$. Indeed, we have $S_N(\mathscr{M}) = sp\{\phi_0,...,\phi_N\}$, the *span* of $\{\phi_0,...,\phi_N\}$.

We shall see below (cf. (3.1.1-2)) that S_N usually operates on a subset $S_M(\mathscr{M})$ of $\mathscr{M}$ with $M \geq N$, typically

$$S_N(\mathscr{M}) \subset S_M(\mathscr{M}) \subset \mathscr{M}.$$

M may be infinite, and it is only in this case that we deal with all of $\mathscr{M}$. We shall use M' to denote $M + 1$ in analogy with to use of $N' = N + 1$. Moreover, only then is our assumption that $\mathscr{M}$ is a Hilbert space relevant.

We stress that while $R_N(\mathscr{M})$ is usually represented by $\mathbb{R}^{N'}$ or by $\mathop{¢}^{N'}$ we may also employ for this purpose spaces $S_N(\mathscr{M})$ that themselves have a ring of functions as a coefficient set (e.g., representations using spline bases). For instance, $R_N(\mathscr{M}) = (\mathbb{R}^p[x])^{N'}$, the space of N'-tuples of polynomials of degree p is of practical interest as indeed, in principle, are most other finite subspaces of the particular function space in question. Actual computation in such a case would depend on an additional isomorphism (so that we could ultimately get at the coefficients), viz

$$i \, i \colon S_N(\mathscr{M}) \rightarrow R_p R_N(\mathscr{M}) = \mathbb{R}^p \times \mathbb{R}^p \times ... \times \mathbb{R}^p, \tag{3.1-4}$$

the latter being an N'-fold direct product. Since the latter contains $N'(p + 1) := N'p'$ degrees of freedom, it is an element of $\mathbb{R}^{N'p'}$. Conceivably and corresponding to the structure of the basis system, this nesting process may be more extensive.

A basis Φ is useful if it has certain properties that make for ease in computation. However, there are also certain necessary requirements concerning the interaction of bases with rounding, algebraic operations, etc. These requirements are best understood in context. Thus here we list some of these properties with references to the places where they are developed below.

Property 1: *Closedness under the operation* $\circ \in \{ + ,-,\bullet,/,\int \}$ *i.e.,*

$$\sum_{i=0}^{N} a_i\phi_i \circ \sum_{j=0}^{N} b_j\phi_j = \sum_{k=0}^{M} c_k\phi_k \qquad (3.1\text{-}5)$$

for all $\circ \in \{ + ,-,\bullet,/,\int \}$. The following condition is typically sufficient for closedness:

$$\phi_i \circ \phi_j = \sum_{k=0}^{M} \gamma_{ijk}\phi_k. \qquad (3.1\text{-}6)$$

In general M will be infinite, but even then a finite M may be used instead if a bound is known for the corresponding error which is thereby introduced.

Property 2: *Consistency with respect to the rounding,* i.e., Eq. (3.1.2-37) below has a solution.

Property 3: *Consistency with respect to relative rounding,* i.e., either (3.1.2-43) holds or alternatively both (5.4-7) and (5.4-8) hold.

Property 4: *Property of inclusion* (cf. (3.2.1)).

Finally we define the approximate operations, namely, the operations defined on the screen. These are induced by the operations in $\mathcal{M}$ and the rounding S_N as we shall now specify.

Consider the structure $(\mathcal{M}; +,-,\bullet,/,\int)$ consisting of the function space $\mathcal{M}$ and the indicated operations. The operations are defined conventionally. The rounding S_N induces a corresponding structure

$$(S_N(\mathcal{M}); \boxplus, \boxminus, \boxdot, \boxslash, \boxed{\int}) \qquad (3.1\text{-}7)$$

in $S_N(\mathcal{M})$. The structure (3.1-7) is given the name *functoid*. The operations are defined by the following property, called *semimorphism:*

$$\bigwedge_{\circ \in \{+,-,\bullet,/,\int\}} \bigwedge_{y,z \in S_N(\mathcal{M})} y \boxed{\circ} z := S_N(y \circ z). \qquad (3.1\text{-}8)$$

For the monadic operator $\int$, y should be deleted from the definition (3.1-8). For computer realization of (3.1-8) see Section 7.1.4.

The isomorphism i induces a structure in $R_N(\mathcal{M})$, which we denote by $(R_N(\mathcal{M}); +,-,*)$. This last structure is simply a finite-dimensional vector space with *scalar product,* the operational symbol for which is denoted by $*$.

Remarks:
1. The sequence of symbols $+,-,\bullet,/,\int$ will appear many times in what follows, and so, it will be convenient to have an abbreviation for this sequence available:

$$\Omega := \{+,-,\bullet,/,\int\}. \qquad (3.1\text{-}9)$$

A corresponding abbreviation $S_N(\Omega)$ is also defined, viz.,

$$S_N(\Omega) = \{\boxplus, \boxminus, \boxdot, \boxslash, \boxed{\int}\}. \qquad (3.1\text{-}10)$$

Thus we have, for example,

$$(\mathcal{M}; +,-,\bullet,/,\int) \equiv (\mathcal{M}; \Omega), \qquad (3.1\text{-}11)$$

and for the functoid in (3.1-7) we may write

$$(S_N(\mathcal{M}); \boxplus, \boxminus, \boxdot, \boxslash, \boxintegral) \equiv (S_N(\mathcal{M}); S_N(\Omega))$$

$$\equiv S_N(\mathcal{M}; \Omega). \qquad (3.1\text{-}12)$$

2. The operations $+, -, \bullet, /, \int$ are basic since other elementary operations such as extracting the kth root, exponentiation, and inverse can be computed iteratively in terms of them. However, for reasons of performance some elementary operations could be adjoined to Ω comprising thereby an enlarged set of "basic" operations. All of our considerations here are valid for any such enlarged set of "basic" operations Ω. Of course all fundamental properties such as (3.1-6) must be maintained for such new operations.

Another perspective of the spaces and operations introduced here is given in Table 3.2-1 below.

3.1.1 Examples of Bases

We give a sampling of bases in Table 3.1.1-1 We stress that our methodology applies more generally.

We include the spline basis and bases of mixed type in Table 3.1.1-1 even though they are a somewhat more general type of basis than the others listed. Many other basis types for computation exist and are indeed used. As with a spline basis, such bases may correspond to discretization methods, finite-element methods, etc.

We are not typically concerned with the values at points of basis elements, but rather only with the behavior of the basis elements in interaction with the operations $\circ \in \{ +, -, \bullet, /, \int, d/dx, ... \}$. Thus in principle, nonexplicitly defined bases (e.g., a basis generated intrinsically by a problem itself) can be of use in our methodology.

3.1.2 Examples of Roundings

As noted in Section 3.1.1, roundings and bases interact. Thus they must be selected so that they are compatible. We make a distinction between

Type and name	basis elements	canonical domains(s)
polynomial type		
Monomial base (MPB)	$1, x, x^2, \dots$	$[-1, 1], [0,1], (-\infty, \infty), [0, \infty)$
Chebyshev base (TPB)	$T_0, T_1, T_2, \dots$	$[-1,1], [0,1]$
Legendre base (LPB)	$P_0, P_1, P_2, \dots$	$[-1,1]$
Bernstein base (BPB)	$B_0, B_1, B_2, \dots$	$[0,1]$
exponential type		
exponential monomial base (EMB)	$1, e^{\alpha x}, e^{2\alpha x}, \dots, \alpha > 0$	$[0, \infty)$
	$1, e^{-\alpha x^2}, e^{-2\alpha x^2}, \dots, \alpha > 0$	$(-\infty, \infty)$
mixed exp/poly base (PEB)	$\{x^m e^{-n\alpha x}\}, \alpha > 0, m, n = 0, 1, \dots$	$[0, \infty)$
Fourier type		
Fourier base (FB)	$\{e^{inx}\}, n = 0, \pm 1, \dots$	$[-\infty, \infty], [0, 2\pi]$
sine base (SB)	$\{\sin nx\}, n = 0, 1, 2, \dots$	$[0, \pi]$
cosine base (CB)	$\{\cos nx\}, n = 0, 1, 2, \dots$	$[0, \pi]$
spline type (discrete or finite element) the coefficient $a_i \in sp\{\phi_j\}$ where the $\{\phi_j\}$ is any base listed in this column. In principle the a_i may be elements of an arbitrary function space; however they typically also fulfill the requirements of any constraints which may be imposed.	$\mathscr{X}_1, \mathscr{X}_2, \dots, \mathscr{X}_N$ $\mathscr{X}_i(x) = \mathscr{X}(x, X_i)$ $:= \begin{cases} 1, x \in X_i \\ 0, oth. \end{cases}$	$[a, b]$ $\bigcup\limits_{i=1}^{n} X_i = [a, b]$ and $\bigwedge\limits_{1 \le i, j \le n} X_i \cap X_j \in \mathbb{R}$
mixed type $\{\phi_i\}$ a basis system in this column. $sp\{\phi_i\} = \{\Sigma_1^n a_i(x)\phi_i(x) \mid a_i \in \mathscr{N}\}$. Coefficient space $\mathscr{N}$ is a space spanned by another basis system in this column $\{\psi_i\}$, say, i.e., $\mathscr{N} = sp\{\psi_i\}$. This mixing of bases may be repeated to provide a nested structure.		

Table 3.1.1-1: Examples of bases

the class of explicit roundings and the class of implicit roundings. In the former case the mapping $S_N: \mathscr{M} \to S_N(\mathscr{M})$ is universal while in the latter case the mapping is problem dependent. Explicit roundings have the advantages of invariance, ease of computation, and simplicity of

error estimates.   The implicit roundings give up these advantages, but
they typically furnish better computational results.

(i) EXPLICITLY DEFINED ROUNDINGS (problem independent)

The result of an operation performed in the functoid $(S_N(\mathcal{M}); \ S_N(\Omega))$
is an element of $\mathcal{M}$ that lies in $S_M(\mathcal{M})$ for some $M \geq N$. (M need not
be finite.)   Thus for the case of linear rounding operators, it suffices to
define $S_N(\phi_k)$ for those basis elements for which $N < k \leq M$:

$$S_N(\phi_k) := \sum_{i=0}^{N} \phi_i a_{ik}, \ \ N < k \leq M. \qquad (3.1.2\text{-}1)$$

The associated $N' \times (M - N)$ matrix $\overset{\wedge}{\mathcal{A}}(S_N) := (a_{ik})_{\substack{0 \leq i \leq N \\ N' \leq k \leq M}}$ will be of
central interest.

Since $S_N(\phi_k) = \phi_k$, $k = 0,...,N$ (compare (3.1-0)), then

$$\mathcal{A}(S_N) = (E, \overset{\wedge}{\mathcal{A}}(S_N)). \qquad (3.1.2\text{-}2)$$

Here E is the $N' \times N'$ identity matrix.   This matrix $\mathcal{A}(S_N)$ may be
viewed as the isomorphic representation of the rounding operator S_N,
i.e.,

$$\mathcal{A}(S_N) := iS_N i^{-1}.$$

Thus

$$S_N(\Phi_M) = \Phi_N * \mathcal{A}(S_N). \qquad (3.1.2\text{-}3)$$

Suppose we have two bases Φ and Ψ such that $sp\Phi_N = sp\Psi_N$ for some
N.   It will be useful to relate the isomorphic representations of the
corresponding rounding operators. Suppose that

$$\Phi_N = \Psi_N * T_N \ \ \text{or} \ \ \Psi_N = \Phi_N * T_N^{-1},$$

where T_N is a matrix. Then for the right member of (3.1.2-3) we have

$$\Phi_N * \mathcal{A} = \Psi_N * T_N * \mathcal{A}.$$

Thus $T_N * \mathscr{A}: sp\Psi_N \rightarrow sp\Phi_N$. Then

$$\Psi_N * T_N * \mathscr{A} * T_N^{-1} =: \Psi_N * \mathscr{B}$$

maps $sp\Psi_N$ into itself. Thus

$$\mathscr{B} := T_N * \mathscr{A} * T_N^{-1}$$

is the isomorphic representation of the rounding operator S_N corresponding to the basis Ψ_N.

One example of the use of this relation is given in Table 3.1.2-1. There
Ψ is the *MPB* and Φ is the *TPB*.

Now let us consider some particular roundings.

Taylor Rounding

Let $\mathscr{M}$ be the set of analytic functions, and let $y \in \mathscr{M}$ have the Taylor
expansion, $y = \sum_{i=0}^{\infty} y_i x^i$. Then $v = S_N y \in S_N(\mathscr{M})$ is the Taylor rounding
of y where

$$y = v + \sum_{i=N'}^{\infty} y_i x^i.$$

The Taylor rounding is realized simply by defining

$$S_N(x^{N+j}) := 0, \quad j \geq 1.$$

Thus referring to (3.1.2-2)

$$\hat{\mathscr{A}}(S_N) = 0.$$

This demonstrates that this rounding is independent of $y(x)$, any particular function to which it is applied. This also displays a formal correspondence between the Taylor rounding and rounding by chopping for
floating-point numbers.

Let $\Phi_N := \{1, x, x^2, ..., x^N\}$ and $X := [-1, 1]$. Then for the rounding error, we have

$$\sigma_j = \| (E - S_N)\phi_j \|_\infty = \| (E - S_N)x^j \|_\infty = \| x^j \|_\infty = 1, \quad j \geq N'.$$

The last follows from the fact that

$$\left.\begin{array}{l} x^{2j+1} \epsilon \; [-1, 1] \\ x^{2j} \quad \epsilon \; [0, 1] \end{array}\right\} x \, \epsilon \, [- 1, 1].$$

This simple observation will be important for developments concerning validation to follow.

Chebyshev Rounding

Let $\mathcal{M}$ be the space of functions having a Fourier-Chebyshev expansion,

$$f(x) = \sum_{i=0}^{\infty}{}' a_i T_i(x).$$

Here $T_i(x)$ is the ith Chebyshev polynomial suitably orthonormalized on $X = [- 1, 1]$, and

$$a_i = \int_{-1}^{1} f(x) T_i(x) \frac{dx}{\sqrt{1-x^2}}, \quad i = 0,1,\dots \; .$$

The prime on the summation symbol means that a_0 is to be halved. The Chebyshev rounding is then given as follows:

$$S_N(T_j(x)) = \begin{cases} T_j(x), & j \leq N \\ 0, & j > N. \end{cases}$$

For rounding errors, we have

$$\sigma_j = \| (E - S_N)T_j(x) \|_\infty = \| T_j(x) \|_\infty = 1, \quad j > N,$$

since $T_j(x) \, \epsilon \, [- 1, 1]$ for $x \, \epsilon \, [- 1, 1]$. We shall use this property in Section 3.2.2.

The Chebyshev rounding can be expressed equivalently in terms of its action on monomials. Indeed, the latter suffices to define the Chebyshev rounding for any polynomial basis (e.g., Legendre and Bernstein). Thus in Table 3.1.2-1 we display the simplest such case. We choose as basis the following multiples of the monomials: $\phi_j = 2^{j-1}x^j$, $j = 0,1,\dots$. We list $S_N(\phi_j)$ and $\sigma_j = \|(E - S_N)\phi_j\|$ for $1 \le N \le 8$, $1 \le j \le 9$ and $j \ge N$, and for the range $X = [-1, 1]$.

Using Table 3.1.2-1, the matrices $\hat{\mathscr{A}}(S_N)$, $N = 1,2,3,\dots$ (cf. (3.1.2-2)) are easily composed:

$$\hat{\mathscr{A}}(S_1) = \begin{pmatrix} 1/2 & 0 & 3/8 & 0 & \cdots \\ 0 & 3/4 & 0 & 10/16 & \cdots \end{pmatrix},$$

$$\hat{\mathscr{A}}(S_2) = \begin{pmatrix} 0 & -1/8 & 0 & -5/32 & \cdots \\ 3/4 & 0 & 10/16 & 0 & \cdots \\ 0 & 8/8 & 0 & 30/32 & \cdots \end{pmatrix},$$

$$\hat{\mathscr{A}}(S_3) = \begin{pmatrix} -1/8 & 0 & -5/32 & \cdots \\ 0 & -5/16 & 0 & \cdots \\ -8/8 & 0 & 30/32 & \cdots \\ 0 & 20/11 & 0 & \cdots \end{pmatrix},$$

$$\cdots$$

Bernstein Rounding

Since the Bernstein polynomials provide a convenient means for verifying the order among functions, as is well known, we include the following observation concerning them.

Let $\mathscr{M}$ be the space of functions have a Fourier-Bernstein expansion. Let $\beta_j^n(x) = x^j(1 - x)^{n-j}$, the jth Bernstein polynomial of degree n. As an example of the use of Table 3.1.2-1, we consider the Chebyshev rounding S_3 of β_4^5:

$$S_3\beta_4^5 = S_3 x^4(1 - x) = S_3(x^4 - x^5).$$

Now using the table for $S_3 x^4$ and $S_3 x^5$, we have

$$S_3\beta_4^5 = (-20x^3 + 16x^2 + 5x - 2)/16.$$

j	ϕ_j	$S_7(\phi_j), S_6(\phi_j)$	σ_j	$S_5(\phi_j), S_4(\phi_j)$	σ_j	$S_3(\phi_j), S_2(\phi_j)$	σ_j	$S_1(\phi_j),$ $S_0(\phi_j)$	σ_j
0	1	1	0	1	0	1	0	1	0
2	$2x^2$	$2x^2$	0	$2x^2$	0	$2x^2$	0	1	1
4	$8x^4$	$8x^4$	0	$8x^4$	0	$8x^2 - 1$	1	3	5
6	$32x^6$	$32x^6$	0	$48x^4 - 19x^2 + 1$	1	$30x^2 - 5$	7	10	22
8	$128x^8$	$256x^6 - 160x^4 + 32x^2 - 1$	1	$224x^4 - 112x^2 + 7$	9	$112x^2 - 21$	37	35	93

j	ϕ_j	$S_8(\phi_j), S_7(\phi_j)$	σ_j	$S_6(\phi_j), S_5(\phi_j)$	σ_j	$S_4(\phi_j), S_3(\phi_j)$	σ_j	$S_2(\phi_j),$ $S_1(\phi_j)$	σ_j	$S_0(\phi_j)$	σ_j
1	x	x	0	x	0	x	0	x	0	0	1
3	$4x^3$	$4x^3$	0	$4x^3$	0	$4x^3$	0	$3x$	1	0	4
5	$16x^5$	$16x^5$	0	$16x^5$	0	$20x^3 - 5x$	1	$10x$	6	0	16
7	$64x^7$	$64x^7$	0	$112x^5 - 56x^3 + 7x$	1	$84x^3 - 28x$	8	$35x$	29	0	64
9	$256x^9$	$576x^7 - 432x^5 + 120x^3 - 9x$	1	$576x^5 - 384x^3 + 54x$	10	$336x^3 - 126x$	46	$126x$	130	0	256

Table 3.1.2–1: Chebyshev rounding of monomials

Upon rearranging the right member here, we may express $S_3\beta_4^5$ in terms of the Bernstein polynomials themselves:

$$S_3\beta_4^5 = (-\beta_5^5 + 18\beta_4^5 + 38\beta_3^5 + 16\beta_2^5 - 5\beta_1^5 - 2\beta_0^5)/16.$$

Spline Rounding

Spline bases differ from other bases in that rounding and the construction of screens is composed of operations on coefficient space(s). The situation is modeled by the framework of finite-element methodology wherein the basic domain X is decomposed into a collection of sets $\{X_i\}_{i=1}^N$. That is, $X = \cup X_i$, where the X_i are disjoint except possibly for sharing portions of boundaries. We shall usually write $\{X_i\}_i$ for $\{X_i\}_{i=1}^N$. In two dimensions the X_i are typically chosen to be triangles. However, in our discussion involving splines we shall for simplicity typically restrict ourselves to one dimension and to the case where the X_i are intervals.

Consider as basis elements the so-called index functions $\mathcal{X}_i(x)$ where

$$\bigwedge_{1 \leq i \leq N} \mathcal{X}_i(x) = \mathcal{X}(x,X_i) := \begin{cases} 1, & x \in X_i \\ 0, & x \notin X_i. \end{cases} \qquad (3.1.2\text{-}4)$$

Here the coefficient space $R(\mathcal{M})$ is taken to be a set $\mathcal{N}$ of N-tuples functions, say $R_N(\mathcal{M}) = \mathcal{N}^N$. Let $\mathcal{BC}$ denote a collection of conditions that these functions satisfy. Consider functions y that have the following form

$$y(x) = \sum_{i=1}^N a_i(x)\mathcal{X}_i \quad \text{with}$$
$$a = (a_i, a_2, \ldots a_N) \in (R_N(\mathcal{M}), \mathcal{BC}). \qquad (3.1.2\text{-}5)$$

The notation $a \in (R_N(\mathcal{M}), \mathcal{BC})$ is taken to mean that the coefficient functions a_i lie in the coefficient space $\mathcal{N}$ and the components satisfy the collection of conditions $\mathcal{BC}$. For conciseness, this collection of functions $y(x)$ in (3.1.2-5) is denoted by $\mathcal{X}^{R_N(\mathcal{M})}$, viz.,

$$\mathscr{X}^{R_N(\mathcal{M})} := sp_{R_N(\mathcal{M})}\{\mathscr{X}_i\}_i := \left\{ \sum_{i=1}^{N} a_i(x)\mathscr{X}_i \mid a_i \in (R_N(\mathcal{M}),\mathscr{B}C) \right\}.$$

(3.1.2-6)

We shall presently make use of another such function $y(x)$. It will be called $z(x)$, and its coefficients will be called $b_i(x)$, $i = 1,...,N$.

By way of an example for $\mathscr{B}C$, let X be an interval in $\mathbb{R}$ and let $\lambda(X_i)$ denote the left and $\rho(X_i)$ denote the right boundary of X_i (so that $\rho(X_i) = \lambda(X_{i+1})$). Then a possible $\mathscr{B}C$ is the following:

$$\mathscr{B}C = \Big\{ \bigwedge_{1 \leq i \leq N} a_i(\rho(X_i)) = a_{i+1}(\lambda(X_{i+1})),$$

$$a_i'(\rho(X_i)) = a'_{i+1}(\lambda(X_{i+1})) \Big\}.$$

If $a \in R_N(\mathcal{M})$ implies that each $a_i \in C_{X_i}^1$, for instance, then $\mathscr{B}C$ here would place y in C_X^1.

It is easily seen that if $(R_N(\mathcal{M}); +,-,\bullet,/,\int)$ is a closed algebraic structure, then so is $(\mathscr{X}^{R_N(\mathcal{M})}; +,-,\bullet,/,\int)$. Indeed, define the algebraic operations in $\mathscr{X}^{R_N(\mathcal{M})}$ as follows:

$$\bigwedge_{\circ \in \{+,-,\bullet,/\}} \bigwedge_{y,z \in \mathscr{X}^{R_N(\mathcal{M})}} y \circ z = \sum_{i=1}^{N} ((a_i \circ b_i)\mathscr{X}_i. \qquad (3.1.2-7)$$

To define integration in $\mathscr{X}^{R_N(\mathcal{M})}$, we restrict our attention to the case where $X = [a,b]$, an interval in $\mathbb{R}$. Let $\{X_i\}_{i=1}^{N}$ be a partition of $[a,b]$ with $X_i \cap X_{i+1} = x_{i+1}$. (For later reference, we call $\xi := x_{k+1}$, with k some arbitrary but fixed value of i.)

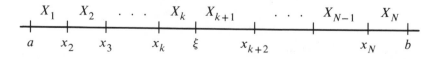

Then with respect to this partition point ξ, the integral of a function

$y \in \mathscr{X}^{-R}{}_N(\mathscr{M})$ is defined as

$$\int_{\xi}^{x} y = \sum_{i=1}^{N} c_i \mathscr{X}_i, \tag{3.1.2-8}$$

where

$$c_i(x) := c_{i-1}(x_i) + \int_{x_i}^{x} a_i(x)dx, \qquad x \in X_i = [x_i, x_{i+1}], \quad i \geq k+2,$$

$$c_{k+1}(x) := \int_{\xi}^{x} a_{k+1}(x)dx, \qquad x \in X_{k+1} = [x_{k+1}, x_{k+2}],$$

$$c_i(x) := c_{i+1}(x_{i+1}) + \int_{x_{i+1}}^{x} a_i(x)dx, \quad x \in X_i = [x_i, x_{i+1}], \quad i \leq k.$$

This restriction to $\mathbb{R}$ and to intervals is made for convenience, but is not essential.

We now note that if $(R_N(\mathscr{M}), \leq)$ is an ordered function space, then so is $(\mathscr{X}^{-R}{}_N(\mathscr{M}), \leq)$, where order for the latter is defined as

$$\bigwedge_{y,z \in \mathscr{X}^{-R}{}_N(\mathscr{M})} y \leq x :\Longleftrightarrow \bigwedge_{1 \leq i \leq N} a_i \leq b_i \Longleftrightarrow a \leq b. \tag{3.1.2-9}$$

To many properties of $R_N(\mathscr{M})$ analogous corresponding properties in $\mathscr{X}^{-R}{}_N(\mathscr{M})$ are induced in just this componentwise manner.

We note in this connection (and for later reference) some properties of spline bases:

$$1 = \sum_{i=1}^{N} \mathscr{X}_i,$$

$$y(x) = y(x) \sum_{i=1}^{N} \mathscr{X}_i =: \sum_{i=1}^{N} y(x) \mathscr{X}_i. \tag{3.1.2-10}$$

Referring to (3.1.2-5), we get

$$y(x) + 1 = \sum_{i=1}^{N} (a_i(x) + 1) \mathscr{X}_i, \tag{3.1.2-11}$$

and for multiplication by a constant or by a function, we get

$$\alpha y(x) = \sum_{i=1}^{N} (\alpha a_i(x)) \mathcal{X}_i.$$

Using (3.1.2-10), we have $\mathcal{M} = \{\sum_{i=1}^{N} y(x) \mathcal{X}_i \mid y \in \mathcal{M}\}$, a virtual tautology in the case that $\mathcal{N} \supset \mathcal{M}$. If the rounding S_N is defined coefficientwise in terms of a rounding S, viz.,

$$S_N(\sum_{i=1}^{N} a_i \mathcal{X}_i) = \sum_{i=1}^{N} S(a_i) \mathcal{X}_i, \qquad (3.1.2\text{-}12)$$

as is natural with spline bases, then we see that

$$R_N(\mathcal{M}) = (S(\mathcal{N}), ..., S(\mathcal{N})) = (S(\mathcal{N}))^N.$$

If $\mathcal{BC}$ contains k conditions for each component a_i, then $S(a_i)$ is a rounding with constraints (cf. (3.1.2-36) and (3.1.2-37) below). Note that $R_N(\mathcal{M})$ itself is a function space and the components a_i of $a \in R_N(\mathcal{M})$ here may themselves be expressed in terms of an expansion. Thus we may generate a nesting process that employs a nest of roundings $S_{N_1} S_{N_2} ...$ (cf. (3.1-4)).

The $\{a_i\}_i$ are independent, and so each may be selected from an independent set of functions, say $\mathcal{N}_i$. Thus by taking $R_N(\mathcal{M}) = \mathcal{N}_1 \times \mathcal{N}_2 \times ... \times \mathcal{N}_N$, we can simply generalize spline representations to

$$\mathcal{X}^{R_N(\mathcal{M})} = \left\{ \sum_{i=1}^{N} a_i(x) \mathcal{X}_i \mid a_i \in \mathcal{N}_i \right\}. \qquad (3.1.2\text{-}13)$$

In analogy with $a \in (R_N(\mathcal{M}), \mathcal{BC})$ we should more properly write $a \in (\mathcal{N} \times ... \times \mathcal{N}_N, \mathcal{BC})$ to denote the constraints imposed on the coefficients to satisfy the relations in $\mathcal{BC}$. An example of $\mathcal{BC}$ in this case is, as before,

$$\mathcal{BC} = \left\{ \bigwedge_{1 \leq i \leq N} \bigwedge_{\substack{a_i \in \mathcal{N}_i \\ a_{i+1} \in \mathcal{N}_{i+1}}} \begin{array}{l} a(\rho(X_i)) = a_{i+1}(\lambda(X_{i+1})), \\ a'_i(\rho(X_i)) = a'_{i+1}(\lambda(X_{i+1})) \end{array} \right\}. \qquad (3.1.2\text{-}14)$$

For convenience and when confusion will not result, we shall hereafter omit the symbol $\mathscr{B}C$

Corresponding to (3.1.2-12), we define

$$S_N(\sum_{i=1}^{N} a_i \mathscr{X}_i) := \sum_{i=1}^{N} S_{N_i,i}(a_i)\mathscr{X}_i, \qquad (3.1.2\text{-}15)$$

where

$$S_{N_i,i}: \mathscr{N}_i \rightarrow S_{N_i,i}(\mathscr{N}_i)$$

and the $S_{N_i,i}(\mathscr{N})$ are independent screens of $\mathscr{M}$.

Thus we see that each component space $\mathscr{N}_i$ and each rounding $S_{N_i,i}(\mathscr{N}_i)$ can be chosen differently and independently for each i. If $\mathscr{B}C$ imposes k_i conditions on the $a_i \in \mathscr{N}_i$ (where, of course, $\sum_{i=1}^{N} k_i$ should correspond to the number of independent conditions found in $\mathscr{B}C$), then $S_{N_i,i}(a_i)$ in (3.1.2-15) is a rounding with constraints (cf. (3.1.2-36) and (3.1.2-37)).

example

We illustrate these concepts of a spline basis with an example:

$$X = [0,\infty], \quad X_1 = [0,1], \quad X_2 = [1,\infty),$$

$$\mathscr{N}_1 = C_{X_1}^0, \quad \mathscr{N}_2 = C_{X_2}^0,$$

$$\phi_1 := \{1,x,x^2\}, \quad \phi_2 := \{1,e^{-x},e^{-2x},e^{-3x}\},$$

$$S_{N_{1,1}}(\mathscr{N}_1) = sp \; \phi_1, \quad S_{N_{2,2}}(\mathscr{N}_2) = sp \; \phi_2.$$

In particular, if y has the form

$$y(x) = a_1(x)\mathscr{X}_1 + a_2(x)\mathscr{X}_2,$$

then the round of y, denoted $S_2 y$, is given by

$$S_2 y(x) = S_{2,1}(a_1)\mathscr{X}_1 + S_{3,2}(a_2)\mathscr{X}_2,$$

where

$$S_{2,1}(a_1) = \alpha_1 + \beta_1 x + \gamma_1 x^2, \qquad\qquad x \in X_1,$$

$$S_{3,2}(a_2) = \alpha_2 + \beta_2 e^{-x} + \gamma_2 e^{-2x} + \delta_2 e^{-3x}, \quad x \in X_2.$$

The coefficients α_1, β_1, γ_1, α_2, β_2, γ_2, δ_2 depend on the explicit mode of rounding, and they become well defined, for example, if the rounding chosen is the Chebyshev rounding.

(ii) IMPLICITLY DEFINED ROUNDINGS (problem dependent)

In this section we employ some notation and properties (such as $\mathcal{A}(\mathcal{L})$, $\mathcal{A}(S_N)$) to be developed with more detail in Chapter 4.

The generic problem in a function space $\mathcal{M}$ may be written

$$\mathcal{L}(y) = 0, \tag{3.1.2-16}$$

where the operator $\mathcal{L}: \mathcal{M} \to \mathcal{M}$. We seek a rounding $S_N: \mathcal{M} \to \mathcal{M}$ so that the solution of

$$S_N \mathcal{L} v = 0 \text{ for } v \in S_N(\mathcal{M}) \tag{3.1.2-17}$$

is equivalent to defining v as the solution of the minimization problem

$$\| \mathcal{L} v \| = \min \tag{3.1.2-18}$$

for a given norm. (Note that v defined by (3.1.2-17) is the Galerkin approximation in $S_N(\mathcal{M})$ to y.) Since (3.1.2-17) is equivalent to $\| S_N \mathcal{L} v \| = 0$, we may combine (3.1.2-17) and (3.1.2-18) to write

$$\| S_N \mathcal{L} v \| = 0 \iff \| \mathcal{L} v \| = \min.$$

Recall that this equivalence defines the rounding S_N implicitly. The norm specified first (possibly dependent upon $\mathcal{L}$) then defines the rounding S_N through (3.1.2-18). The Riesz-Fisher theorem allows us to express the functional (3.1.2-18) of $\mathcal{L} v$ as an inner product. Then (3.1.2-18) may be written

$$< w, \mathcal{L} v > \ = \min.$$

We remark here that this implicit approach to the rounding operator S_N makes for increased difficulty in the ability to determine effective inclusions (i.e., including sets) since S_N depends on $\mathscr{L}$. However, the ability to determine inclusions is not precluded by this implicit approach. The development of such inclusions is novel for Galerkin approximations (cf. (3.2.2)).

In principle it is possible to fix the rounding first in the implicit case, and to let the norm be determined. It is instructive to consider an example of such an approach wherein a familiar rounding is used.

Least-square Rounding

Given $u \in C^0[-1, 1]$ and given a linear manifold $S_N(\mathscr{M})$, the rounding of u, i.e., the function $v \in S_N(\mathscr{M})$ for which $S_N u := v$, is determined by the minimization process

$$\int_{-1}^{1} (v - u)^2 dx = \min.$$

This generates the L^2-norm

$$\| u \|^2 := \, < u, u > \, = \int_{-1}^{1} uu \, dx.$$

The problem to be solved is $\mathscr{L}y = 0$, and for $v \in S_N(\mathscr{M})$, we have

$$\| \mathscr{L}y - \mathscr{L}v \| = \| \mathscr{L}v \| = \min.$$

This conforms to the requirement (3.1.2-18). We write this minimization as

$$< \mathscr{L}v, \mathscr{L}v > \, = \int_{-1}^{1} \mathscr{L}v \mathscr{L}v \, dx = \min. \qquad (3.1.2\text{-}19)$$

We proceed further with this example and seek to solve (3.1.2-19). We consider $v \in sp\{\phi_1, ..., \phi_N\} = S_N(\mathscr{M})$. That is, we consider the representation

$$v = \sum_{i=1}^{N} a_i \phi_i.$$

We take the linear case: $\mathscr{L}y = \ell y - r$, so that the necessary condition for a minimum

$$\frac{\partial}{\partial a_i} < \mathcal{L}v, \mathcal{L}v > \; = 0, \quad 1 \le i \le N,$$

may be written as

$$< \mathcal{L}v, \ell\phi_i > \; = 0, \quad 1 \le i \le N. \qquad (3.1.2\text{-}20)$$

These are customarily referred to as the normal equations. Set $\psi_i = \ell\phi_i$, $1 \le i \le N$, and consider the problem dependent subspace

$$\{\ell f \mid f \in S_N(\mathcal{M})\} = sp\{\psi_i \mid 1 \le i \le N\}.$$

The $\psi_i \in \ell S_N(\mathcal{M})$, $1 \le i \le N$. Then (3.1.2-20) may be written equivalently as follows

$$\bigwedge_{\psi_i \in \ell S_N(\mathcal{M})} < \mathcal{L}v, \psi_i > \; = 0. \qquad (3.1.2\text{-}21)$$

(3.1.2-20) and (3.1.2-21) may be rewritten

$$\sum_{i=1}^{N} a_i < \ell\phi_i, \ell\phi_j > \; - \; < r, \ell\phi_j > \; = 0.$$

Then to approximate the solution of $\mathcal{L}y = 0$, it suffices to compute the entries of the matrix $\Gamma = (\gamma_{ij})$, $\gamma_{ij} = <\ell\phi_i, \ell\phi_j>$, $i,j = 1,...,N$ and the vector $R = (\rho_j)$, $\rho_j = <r, \ell\phi_j>$, $j = 1,...,N$, and then to solve the following linear system for $t = (a_1,...,a_N)^T$:

$$\Gamma t = R. \qquad (3.1.2\text{-}22)$$

example

$$\mathcal{L}y = 1 - y + \int_0^x y, \quad \ell y = -y + \int_0^x y, \quad r = -1.$$
$$\phi_i = x^i, \quad i = 0,1,..., \quad X = [-1, 1].$$
$$S_N(\mathcal{M}) = sp\{1, x, x^2,...,x^N\}.$$

Then

$$\{\ell\phi_i \mid i = 0,1,...\} = \{x - 1, \tfrac{1}{2}x^2 - x, \tfrac{1}{3}x^3 - x^2,...\}.$$

Also

$$<x^i,x^j> = \begin{cases} \dfrac{2}{i+j+1}, & i+j \text{ even,} \\[2mm] 0, & i+j \text{ odd.} \end{cases}$$

For this case $N = 3$, $v_3 := a + bx + cx^2$ and

$$\Gamma = \begin{pmatrix} \dfrac{8}{3} & -1 & \dfrac{12}{15} \\[2mm] -1 & \dfrac{46}{60} & -\dfrac{1}{3} \\[2mm] \dfrac{12}{15} & -\dfrac{1}{3} & \dfrac{136}{315} \end{pmatrix}, \quad R = \begin{pmatrix} 2 \\[2mm] -\dfrac{1}{3} \\[2mm] \dfrac{2}{3} \end{pmatrix}.$$

In this case, using (3.1.2-22) yields

$$v_3(x) = 1.004 + 1.1099x + 0.5406x^2.$$

Correspondingly in the case $N = 2$ we get

$$v_2(x) = 1.149 + 1.064x.$$

Note that $y(1) = e$ (the Euler constant) and that $v_3(1) = 2.6545$.

Now let us turn the example around and determine the rounding operator S_2 so that the condition

$$S_2\mathscr{L}v = 0 \iff \|\mathscr{L}v\| = \min \qquad (3.1.2\text{-}23)$$

holds (compare (3.1.2-17) and (3.1.2-18)f).

Let $\Phi = (1,x,x^2,...)$, $\tilde{t} = (a,b,c,0,0,...)$, $s = (s_0,s_1,s_2,...)$, and write $v = \Phi*\tilde{t}$ and $r = \Phi*s$. Then

$$\mathscr{L}v = \mathscr{L}(\Phi*\tilde{t}) = \Phi*\mathscr{A}(\ell)\tilde{t} - \Phi*s,$$

$\mathcal{A}(\ell)$ is an appropriate matrix, see (4.1-1)f. In the right member here, t is a column vector (of three elements), and $\mathcal{A}(\ell)$ is a matrix with three columns (which are in general infinitely long). Alternatively we have

$$\mathcal{L}v = \Phi*(s,\mathcal{A}(\ell))(1, \widetilde{t})^T = \Phi*\mathcal{A}(\mathcal{L})t, \qquad (3.1.2\text{-}24)$$

where $\mathcal{A}(\mathcal{L}) = (s,\mathcal{A}(\ell))$ and $t := (1, \widetilde{t})^T$, both blockwise.

Now let

$$w = \Phi*u$$

be a generic element of $\mathcal{M}$. Then

$$S_2 w = S_2\Phi*u = S_2(\Phi)*u = \phi*\mathcal{A}(S_2)u. \qquad (3.1.2\text{-}25)$$

Here $\mathcal{A}(S_2)$ the matrix that we seek, consists of three rows that are infinitely long.

Combining (3.1.2-23) - (3.1.2-25), we obtain

$$0 = S_2\mathcal{L}v = S_2\Phi*\mathcal{A}(\mathcal{L})t = \Phi*\mathcal{A}(S_2)\mathcal{A}(\mathcal{L})t.$$

From this in turn we have equivalently

$$\mathcal{A}(S_2)\mathcal{A}(\mathcal{L})t = 0. \qquad (3.1.2\text{-}26)$$

Equation (3.1.2-22), equivalent to the minimization problem $\|\mathcal{L}v\| = \min$, may be written

$$(-R, \Gamma)t = 0. \qquad (3.1.2\text{-}27)$$

Combining (3.1.2-26) and (3.1.2-27), we get

$$\mathcal{A}(S_2)\mathcal{A}(\mathcal{L}) = (-R, \Gamma) = \begin{vmatrix} -2 & \dfrac{8}{3} & -1 & \dfrac{4}{5} \\[2mm] \dfrac{1}{3} & -1 & \dfrac{46}{60} & -\dfrac{1}{3} \\[2mm] -\dfrac{2}{3} & \dfrac{4}{5} & -\dfrac{1}{3} & \dfrac{136}{315} \end{vmatrix}, \qquad (3.1.2\text{-}28)$$

the last member here following from the explicit example just treated.

Using $v = a + bx + cx^2$, we see that

$$\mathscr{L}v = 1 - a + (a - b)x + (\tfrac{1}{2}b - c)x^2 + \tfrac{1}{3}cx^3.$$

Then from (3.1.2-24) we have

$$\mathscr{A}(\mathscr{L}) = \begin{pmatrix} 1 & -1 & 0 & 0 \\ 0 & 1 & -1 & 0 \\ 0 & 0 & \dfrac{1}{2} & -1 \\ 0 & 0 & 0 & \dfrac{1}{3} \end{pmatrix}.$$

Indeed, we expect the inverse of $\mathscr{A}(\mathscr{L})$ to exist since $\mathscr{L}$ is not degenerate:

$$\mathscr{A}(\mathscr{L})^{-1} = \begin{pmatrix} 1 & 1 & 2 & 6 \\ 0 & 1 & 2 & 6 \\ 0 & 0 & 2 & 6 \\ 0 & 0 & 0 & 3 \end{pmatrix}.$$

Using this and (3.1.2-28), we compute

$$\mathscr{A}(S_2) = (-R, \Gamma)\mathscr{A}(\mathscr{L})^{-1} = \begin{pmatrix} -2 & \dfrac{2}{3} & -\dfrac{2}{3} & \dfrac{2}{5} \\ \dfrac{1}{3} & -\dfrac{2}{3} & \dfrac{1}{5} & -\dfrac{3}{5} \\ -\dfrac{2}{3} & \dfrac{2}{15} & -\dfrac{2}{5} & \dfrac{2}{21} \end{pmatrix},$$

Now using (3.1.2-25), we can calculate in particular,

$S_2(x^i)$, $i = 0,1,2,3$:

$$S_2(1) = (1,x,x^2,\ldots)*\mathscr{A}(S_2)*(1,0,0,\ldots)^T$$

$$= -2 + \frac{1}{3}x - \frac{2}{3}x^2,$$

$$S_2(x) = \frac{2}{3} - \frac{2}{3}x + \frac{2}{15}x^2, \qquad (3.1.2\text{-}29)$$

$$S_2(x^2) = -\frac{2}{3} + \frac{1}{5}x - \frac{2}{5}x^2,$$

$$S_2(x^3) = \frac{2}{5} - \frac{3}{5}x + \frac{2}{21}x^2.$$

This rounding S_2 is defined on $[-1, 1]$, and as the derivation here shows, it is dependent on $\mathscr{L}$, i.e., it is problem dependent. Typically, such roundings do not satisfy property (3.1-0). The computation of $\mathscr{A}(S_2)$ is necessary for validation (cf. (3.2.2(ii)). A simplistic approach to this question would be to define the directed rounding $IS_2(x^i) = S_2(x^i) + [-1, 1]\sigma_i$ in terms of the particular rounding error $\sigma_i = \|S_2(x^i) - x^i\|$. While this approach is more properly associated with the explicit case, we remark here that it is central for obtaining inclusions in all cases of implicit roundings corresponding to spectral methods and pseudospectral methods, e.g., collocation methods.

We interpolate a remark about the normal equations (3.1.2-20). If the basis $\{\phi_i\}$ is chosen so that the functions $\{\psi_i = \ell\phi_i\}$ form an orthonormal system, then the normal equations define their solution explicitly as

$$a_i = \langle r,\psi_i \rangle, \quad 1 \le i \le N.$$

Ritz–Galerkin Rounding

From (3.1.2-21) we see that implicit roundings are definable through a process of Galerkin approximation (in fact, Ritz-Galerkin) with respect to a set of test functions. It shall be instructive to review our previous process in this light since we shall then see a correspondence that exists between the set of test functions and the rounding operator.

Let $\mathscr{E}$ be a complete set of test functions, i.e., $\mathscr{E} \subset \mathscr{M}$, such that for any $f \in \mathscr{M}$

$$\bigwedge_{\psi \in \mathscr{E}} \int_{-1}^{1} f(t)\psi(t)dt = <f,\psi> = 0 \Rightarrow f = 0. \qquad (3.1.2-30)$$

Now select a subset $S_N(\mathscr{E}) = \ell S_N(\mathscr{M})$ and interpret (3.1.2-21) as a finite test for $\mathscr{L}v = 0$. In particular, choose $S_N(\mathscr{E}) = \{\psi_1,...,\psi_N\}$, and make the N tests:

$$\bigwedge_{1 \leq i \leq N} <\mathscr{L}v,\psi_i> = 0. \qquad (3.1.2-31)$$

Then

$$<\ell v, \psi_j> = <r,\psi_j>, \quad 1 \leq j \leq N,$$

or

$$<\Phi * \mathscr{A}(\ell)\widetilde{t},\psi_j> = <r,\psi_j>, \quad 1 \leq j \leq N,$$

where $\mathscr{A}(\ell)$ is an approximate matrix. In terms of the matrix $\mathscr{B} = (\mathscr{B}_{ji}) := (< \phi_i,\psi_j >)$ and the vector $R := (< r,\psi_j >)$, this last relation is more compactly written

$$\mathscr{B}\mathscr{A}(\ell)\widetilde{t} = R. \qquad (3.1.2-32)$$

Let us now return to our example; to the basis $\Phi = (1,x,x^2,...)$ and to the test system $\Psi = \{1,x,x^2,...\}$. In this case $(N = 3)$ we have

$$\mathscr{B} = \begin{pmatrix} 2 & 0 & \frac{2}{3} & 0 & . & . & . \\ 0 & \frac{2}{3} & 0 & \frac{2}{5} & . & . & . \\ \frac{2}{3} & 0 & \frac{2}{5} & 0 & . & . & . \end{pmatrix}, \quad R = \begin{pmatrix} -2 \\ 0 \\ -\frac{2}{5} \end{pmatrix}$$

and

$$\mathscr{A}(\ell) = \begin{pmatrix} -1 & 0 & 0 \\ 1 & -1 & 0 \\ 0 & \dfrac{1}{2} & -1 \\ 0 & 0 & \dfrac{1}{3} \\ 0 & 0 & 0 \\ & \cdots & \end{pmatrix}.$$

Setting $\tilde{t} = (a,b,c,0,0,...)^T$ and $\Gamma = \mathscr{B}\mathscr{A}(\ell)$, (3.1.2-32) becomes $\Gamma\tilde{t} = R$, or explicitly,

$$\begin{pmatrix} -2 & \dfrac{1}{3} & -\dfrac{2}{3} \\ \dfrac{2}{3} & -\dfrac{2}{3} & \dfrac{2}{15} \\ -\dfrac{2}{3} & \dfrac{1}{5} & -\dfrac{2}{5} \end{pmatrix} \begin{pmatrix} a \\ b \\ c \end{pmatrix} = \begin{pmatrix} -2 \\ 0 \\ -\dfrac{2}{3} \end{pmatrix}. \qquad (3.1.2\text{-}33)$$

Then $\tilde{t} = (1, \dfrac{10}{9}, \dfrac{5}{9})$. Then we find that $v_2(x) = 1 + \dfrac{10}{9}x + \dfrac{5}{9}x^2$ and $e \sim v(1) = \dfrac{24}{9} = 2.666$.

Writing (3.1.2-33) as $(-R, \Gamma)t = 0$ (in analogy with (3.2.2-27)), we similarly derive the following relation for $\mathscr{A}(S_2)$:

$$\mathscr{A}(S_2) = (-R, \Gamma)\mathscr{A}(\mathscr{L})^{-1}, \qquad (3.1.2\text{-}34)$$

where $\mathscr{A}(\mathscr{L})^{-1}$ is the inverse of the matrix appearing in (3.1.2-24). In the case of the example,

$$\mathcal{A}(S_2) = \begin{pmatrix} 2 & 0 & \dfrac{2}{3} & 0 \\[8pt] 0 & \dfrac{2}{3} & 0 & \dfrac{2}{5} \\[8pt] \dfrac{2}{3} & 0 & \dfrac{2}{5} & 0 \end{pmatrix},$$

which, not surprisingly, is the leading part of the matrix $\mathcal{B}$ itself.

Thus we see as previewed that the rounding through its matrix representation $\mathcal{A}(S_2)$ is determinable from the basis and test set. We also see that we may obtain explicit roundings as implicitly defined roundings by means of special choices of the set of test functions $S_N(\mathcal{E})$.

In the present setting, however, we may consider more tests in (3.1.2-31) than there are unknowns. Say we set $v = \sum_{i=1}^{N} a_i \phi_i$ and require that $< \mathcal{L}v, \psi_j > \; = 0$ for $\psi_j \in S_M(\mathcal{E})$, $i \le j \le M, M > N$. The resulting overdetermined system can be solved by least squares.

For instance, setting $\Gamma = \mathcal{B}\mathcal{A}(\ell)$, an $M \times N$ matrix, we solve $\Gamma \tilde{t} = R$ through

$$\Gamma^T \Gamma \tilde{t} = \Gamma^T R. \tag{3.1.2-35}$$

This equation may be written equivalently as

$$(-\Gamma^T R, \; \Gamma^T \Gamma)t = 0.$$

Then

$$\mathcal{A}(S_2) = (-\Gamma R, \; \Gamma^T \Gamma)\mathcal{A}(\mathcal{L})^{-1} = \Gamma^T(-R, \; \Gamma)\mathcal{A}(\mathcal{L})^{-1}.$$

example

We return to our example for $M = 4$, $N = 3$. Then

$$
\mathcal{B} = \begin{pmatrix}
2 & 0 & \frac{2}{3} & 0 & . & . & . \\
0 & \frac{2}{3} & 0 & \frac{2}{5} & . & . & . \\
\frac{2}{3} & 0 & \frac{2}{5} & 0 & . & . & . \\
0 & \frac{2}{5} & 0 & \frac{2}{7} & . & . & .
\end{pmatrix}, \quad
R = \begin{pmatrix}
-2 \\
0 \\
-\frac{2}{3} \\
0
\end{pmatrix},
$$

$$
\Gamma = \mathcal{B}\mathcal{A}(\ell) = \begin{pmatrix}
-2 & \frac{1}{3} & -\frac{2}{3} \\
\frac{2}{3} & -\frac{2}{3} & \frac{2}{15} \\
-\frac{2}{3} & \frac{1}{5} & -\frac{2}{5} \\
\frac{2}{5} & -\frac{2}{5} & \frac{2}{21}
\end{pmatrix},
$$

$$
\Gamma^T\Gamma = \begin{pmatrix}
5.04\overline{8} & -1.40\overline{4} & 1.7269841 \\
-1.40\overline{4} & 0.7\overline{5} & -0.42920635 \\
1.7269841 & -0.4290635 & 0.63129252
\end{pmatrix},
$$

$$
\Gamma^T R = \begin{pmatrix}
\frac{40}{9} \\
-\frac{4}{5} \\
\frac{8}{5}
\end{pmatrix}, \quad
\tilde{t} = \begin{pmatrix}
a \\
b \\
c
\end{pmatrix} = \begin{pmatrix}
1.0009014 \\
1.1175076 \\
0.55616133
\end{pmatrix}.
$$

Then

$$v(x) = a + bx + cx^2,$$

and

$$e \sim v(1) = 2.6745703.$$

Here

$$\mathcal{A}(S_2) = \begin{pmatrix} -4.\overline{4} & 0.60\overline{4} & -1.6 & 0.3809523 \\ 0.8 & -0.60\overline{4} & 0.30\overline{2} & -0.3809523 \\ -1.6 & 0.1269841 & -0.6041588 & 0.081401 \end{pmatrix}.$$

From the columns of $\mathcal{A}(S_2)$ we may read-off the action of S_2 on the monomials:

$$S_2(x^0) = -4.\overline{4} + 0.8x - 1.6x^2,$$

$$S_2(x^1) = 0.60\overline{4} - 0.60\overline{4}x + 0.1269841x^2,$$

$$S_2(x^2) = -1.6 + 0.30\overline{2}x - 0.6041588x^2,$$

$$S_2(x^3) = 0.3809523 - 0.3809523x + 0.081401x^2.$$

Optimal Implicit Roundings

Here corresponding to the given problem $\mathcal{L}Y = 0$ and a given integer N, we define an optimal implicit rounding $S_N^{|\cdot|_\infty}$, as follows

$$S_N^{|\cdot|_\infty}\mathcal{L}v = 0 :\Longleftrightarrow \|y - v\|_\infty = \min.$$

In Section 3.2.(ii) we consider inclusions, and we shall see that there is a correspondence between the diameter of an optimal inclusion and the value of the minimum associated with the definition of $S_N^{|\cdot|_\infty}$ here.

(iii)   ROUNDINGS WITH CONSTRAINTS

Let $c(y) = 0$ denote (a set) of constraints, such as, for example:

$$y(0) - 1$$

$$c(y) = \quad y'(1) + 2y(0) - 3 \quad .$$

$$\int_0^1 y \, dt$$

Then a rounding operator $S_N: \mathcal{M} \to S_N(\mathcal{M})$ is said to fulfill a constraint c if

$$\bigwedge_{y \in \mathcal{M}} c(S_N(y)) = 0.$$

To construct a rounding S_N that satisfies this requirement, we start with a rounding $\tilde{S}_N$ that has the property $\tilde{S}_N(\mathcal{M}) \supset S_N(\mathcal{M})$. Then we impose the conditions of constraint on $\tilde{S}_N$, particularizing it until we arrive at S_N. We employ the basis $\Phi_N = \{\phi_0,...,\phi_N\}$, and we suppose that $c(y)$ is composed of k linear independent constraints. S_N is sought in the form

$$S_N y = \tilde{S}_{N-k} y + \sum_{j=1}^{k} \alpha_j \phi_{N-k+j}, \qquad (3.1.2\text{-}36)$$

where the coefficients $\alpha_0,...,\alpha_k$ are determined by solving the following linear system

$$c\left(\tilde{S}_{N-k} y + \sum_{j=1}^{k} \alpha_j \phi_{N-k+j} \right) = 0. \qquad (3.1.2\text{-}37)$$

If the constraints are not linear, we shall be unable in general to solve (3.1.2-37) explicitly so that the rounding S_N is implicitly defined. This construction partitions $sp\Phi$ into two parts $sp\{\phi_{N-k+j}\}_{j=1}^{k}$ and its complement in $sp\Phi$ with $\tilde{S}_{N-k}$ being a rounding onto the latter. In many cases this literal form of the partition of $sp\Phi$ causes the system (3.1.2-37) to be singular. A permutation of the basis frequently remedies this difficulty. The permutation allows us to choose any $N' - k$ of the basis functions in Φ as a basis for the image of $\widetilde{S}_{N-k}$ (not simply the first $N' - k$ as the notation in (3.1.2-37) implies).

Examples of function space problems in which constraints appear are
(a) differential equations where the regular boundary conditions may be interpreted as constraints;
(b) spline bases with boundary conditions, cf. Section 4.2 below;
(c) problems with a manifold of solutions (undetermined problems) for which constraints are added to specify a unique solution.

We now give several examples of the rounding construction process that we have described in this section.

examples

(1) $c(y) :\equiv y(0) = 0, \quad \Phi = \{1,x,...\}.$

The system (3.1.2-37) would yield

$$c(\widetilde{S}_{N-1}y + \alpha x^N) = 0,$$

which gives no condition on α at all, i.e., the system is singular. Interchanging $\phi_0 = 1$ and $\phi_N = x^N$ results in the rounding

$$S_N y := x\overset{\approx}{S}_{N-1}y,$$

where the range of $\overset{\approx}{S}_{N-1}$ is $sp\{1,...,x^{N-1}\}$. For consistency with our previous notation, we write $x\overset{\approx}{S}_{N-1} = \widetilde{S}_{N-1}$, so that by a new identification

$$S_N y = \widetilde{S}_{(N-1)}y.$$

$\widetilde{S}_{N-1}$ is a rounding onto $sp\{x,...,x^N\}$.

(2) $c(y) := (y(0) - \eta, y(1) - \theta)^T, \quad \Phi = \{1,x,...\},$

$$S_N y := \eta(1-x) + \theta x + x(1-x)\overset{\approx}{S}_{N-2}y.$$

As in example (1), we could write $x(1-x)\overset{\approx}{S}_{N-2} = \widetilde{S}_{N-2}$ for notational consistency.

(3) $c(y) := (y(\infty), \int_0^\infty y dt - 1)^T, \quad \Phi = \{1, e^{-x}, e^{-2x}, ...\},$

$$S_N y := e^{-x} \overset{\approx}{S}_{N-2} y - N(\int_0^\infty e^{-t} \overset{\approx}{S}_{N-2} y dt - 1) e^{-Nx}.$$

(iv) RELATIVE ROUNDINGS

In Chapter 5 we shall introduce iterative residue correction for which a so-called relative rounding is required. This type of rounding is easily modeled in the context of power bases such as *MPB*, *EMB*, *PEB* listed in Section 3.1.1.

A power base $\Phi = \{\phi_0, \phi_1, ...\}$ has the following important property:

$$\sum_{i=k}^{M} a_i \phi_i = \left(\sum_{i=0}^{M-k} a_{i+k} \phi_i \right) \phi_k, \qquad (3.1.2\text{-}38)$$

which may be viewed informally as a scaling. We introduce some notation in order to have a uniquely defined isomorphic representation in the coefficient space. Let

$$y = \sum_{i=k}^{M} a_i \phi_i =: \Phi * \underset{k \text{ zeros}}{(0, ..., 0, a)}, \qquad (3.1.2\text{-}39)$$

where $a = (a_k, ..., a_M)$, and

$$y = \left(\sum_{i=0}^{M-k} a_{i+k} \phi_i \right) \phi_k. \qquad (3.1.2\text{-}40)$$

Then

$$iy = \underset{k \text{ zeros}}{(0, ..., 0, a_k, a_{k+1}, ..., a_M)}$$

represents y isomorphically. We call

$$m(y) = \sum_{k=0}^{M-k} a_{i+k} \phi_i \qquad (3.1.2\text{-}41)$$

the mantissa of y, and and we call

$$sf(y) = \phi_k \qquad\qquad (3.1.2\text{-}42)$$

the shifting factor of y. Correspondingly, we call

$$im(y) := a$$

the isomorphic mantissa of y, and we call

$$isf(y) := k$$

the isomorphic shifting factor of y.

For power bases with these constructs supplied, a *relative rounding* S_N^p with a *scaling index* p is defined as follows: Let

$$y = \sum_{i=0}^{M} a_i \phi_i.$$

Then

$$S_N^p y := \phi_p S_N \sum_{j=0}^{M-p} a_{j+p} \phi_j. \qquad\qquad (3.1.2\text{-}43)$$

Unlike other roundings that are mappings onto the "leading part", the relative rounding S_N^p maps onto some "middle part". The reasonableness of this seemingly curious rounding will be shown in Chapter 5 where S_N^p is applied and plays a central role in iterative correction processes. Note that with scaling index $p = 0$, $S_N^0 \equiv S_N$. In terms of the other constructs introduced here, note that ϕ_p is the shifting factor of $w := \phi_p S_N \sum_{j=0}^{M-p} a_{j+p} \phi_j$ so that p is its isomorphic shifting factor. The constructs introduced here have their counterparts in nonpower bases. See Section 5.4. With these conventions for relative rounding, the screen $S_N(\mathcal{M})$ now denotes an enlarged set of functions. For example

$$S_N \mathbb{R}[x](X) = \{y \mid y = S_N^p z, \, z \in \mathbb{R}[x](X), \, p \in \mathbf{N} \cup \{0\}\}.$$

3.1.3 Operations and Iteration

We conclude Section 3.1 with some observations concerning operations and iterations in $S_N(\mathcal{M})$. These observations will be enlarged upon in Section 4.2 and in Chapter 7.

(i) In order to satisfy the property (3.1-8) it is necessary to compute best approximations in $S_N(\mathcal{M})$ to elements $h \in \mathcal{M}$. When S_N is defined through generalized-Fourier series, as in some of our examples, best is taken to mean with respect to the norm in $\mathcal{M}$. Conceivably other norms may be used.

(ii) The iteration process (cf. (2.1-4))

$$y_{i+1} = (y_i), \quad i \geq 0, \tag{3.1.3-1}$$

with y_0 given, is transformed into an iteration process in $S_N(\mathcal{M})$, which we write

$$v_{i+1} = \tilde{f}(v_i), \quad i \geq 0, \tag{3.1.3-2}$$

with v_0 given. Later on, in Chapter 7, (3.1.3-2) will be referred to as the computer-screen implementation of (3.1.3-1). Three possible methods for implementing $\tilde{f}$ are discussed in Section 7.1.

Common practice dictates that initial iterates need not be computed to full accuracy. It suffices to have methods that successively refine more terminal approximations in order that as accurate a result as possible can be obtained within the arithmetic setting in use. When such terminal-type correction methods are possible it is clearly economical to save complexity in computing the rounding S_N (i.e., the approximation of f by $\tilde{f}$) and to use up the complexity in the final steps.

3.2 SPACES, BASES, AND ROUNDINGS FOR VALIDATION

We now introduce some so-called higher spaces, which are required for the validation of solutions to problems set in $\mathcal{M}$, in particular, for the computation of sets, which include such solutions.

$I\!I\mathbb{R}^d$ denotes the *space of interval subsets* of $\mathbb{R}^d$, i.e., intervals of sets of d-dimensional vectors.

We now review some commonly used notation concerning interval spaces. Let $\mathcal{N}$ be a Banach space, and let $P\mathcal{N}$ be its power set. The set of all convex, closed, and bounded subsets of $\mathcal{N}$ will be denoted by $I\!L\mathcal{N}$. Clearly, $I\!L\mathcal{N} \subset P\mathcal{N}$. $I\!L\mathcal{N}$ is a set of *generalized intervals*. If $\{\mathcal{N}, \le\}$ is an ordered Banach space in the graph sense (cf. (2.2-3)), we simplify the set $I\!L\mathcal{N}$ by taking it to be composed of the elements called intervals of the form $F = [f,g] \in I\!L\mathcal{N}$ where

$$F = \{y \mid f \le y \le g\},$$

provided that $f \le g$. Typically the order relation is pointwise. In practice and in our exposition to follow, we shall restrict the choice of endpoints f and g of the interval F to a subset $\tilde{\mathcal{N}}$ of $\mathcal{N}$. The collection of all such corresponding intervals is denoted by $I\!L\tilde{\mathcal{N}}$. For example, if $\mathcal{N} = C^0(X)$, a choice for $\tilde{\mathcal{N}}$ is $\mathbb{R}[x](X)$ (the collection of polynomials in x) and $I\!I\mathbb{R}[x](X)$ is the set of so-called *interval polynomials* [7].

The functional $d: I\!L\mathcal{N} \to \mathbb{R}$ is called the *diameter*, viz.,

$$d(F) := g - f.$$

Now take $\mathcal{N}$ to be our function space $\mathcal{M}$, which is assumed to be a separable Hilbert space. Consider the set of linear combinations of the basis $\{\phi_i\}_{i=0}^{\infty} \subset \mathcal{M}$ taken with interval coefficients. That is, $\sum_{i=0}^{\infty} A_i\phi_i$ where $A := (A_0, A_1,...)$ is a sequence whose elements are chosen from any interval space such as $I\!I\mathbb{R}$, $I\!I\mathbb{R}[x](X),...$. The finite span $Isp_{I\!I\mathbb{R}}\Phi = \sum_{i=0}^{N} A_i\phi_i$ of such objects, denoted by $I\!Isp\{\phi_i\}_{i=0}^{N}$, is the counterpart to the set $I\!I\tilde{\mathcal{M}}$ in the notation above.

For each $\xi \in X$, the symbol $\sum_{i=0}^{N} A_i\phi_i(\xi)$ denotes a linear combination of the intervals A_i, $i = 0,..., N$. However, as we shall see, $\sum_{i=0}^{N} A_i\phi_i$ may be interpreted as a set of functions, that is, as an element of $P\mathcal{M}$. Then the $I\!Isp\{\phi_i\}_{i=0}^{N}$ belongs to the power set $P\mathcal{M}$ of $\mathcal{M}$ as well. It is called a *set-screen* of $P\mathcal{M}$ and is denoted by $IS_N(P\mathcal{M})$. (The dual nature of the symbol $\sum_{i=0}^{N} A_i\phi_i$ will play a key role in the implementation of interval ultra-arithmetic. See Remark 7.1.4-1.) The operator

$$IS_N\colon \boldsymbol{P\mathcal{M}} \to \boldsymbol{II}\,sp\{\phi_i\}_{i=0}^{N}$$

is the *directed projection* or (including nonlinear roundings) it is the
directed rounding onto the set-screen $IS_N(\boldsymbol{P\mathcal{M}})$. Indeed, IS_N has the
following property (*directed rounding*):

$$\bigwedge_{f\in\mathcal{M}} f \in IS_N f,$$

$$\bigwedge_{F\in\boldsymbol{P\mathcal{M}}} F \subset IS_N F, \qquad\qquad (3.2\text{-}1)$$

$$\bigwedge_{G\in IS_N(\boldsymbol{P\mathcal{M}})} G = IS_N G.$$

Here containment (i.e., the symbols $\in$ and $\subset$) is meant in the graph
sense. For instance, $IS_N f$ is a set of functions and for each $x \in X$,
$IS_N f(x)$ is some interval (a point set). Then $f \in IS_N f$ means that
$f(x) \in IS_N f(x)$ for each $x \in X$. $\subset$ has a corresponding meaning. Of
course, this containment is exactly the one discussed in the beginning of
Section 2.2. Likewise $\sum_{i=0}^{\infty} A_i\phi_i$ is that set of functions each of whose
graphs is contained in the graph of $\sum_{i=0}^{\infty} A_i\phi_i$. Among these functions are
$\{\sum_{i=0}^{\infty} a_i\phi_i \mid a_i \in A_i, i = 0,..., N\}$. The boundary of the set $\sum_{i=0}^{N} A_i\phi_i$ is in
general composed piecewise of the functions ϕ_i, $i = 0,..., N$.

For computer representation, we introduce an isomorphism i,

$$i\colon IS_N(\boldsymbol{P\mathcal{M}}) \to IR_N(\boldsymbol{P\mathcal{M}}),$$

where $IR_N = iIS_N$. The isomorphic space $IR_N(\boldsymbol{P\mathcal{M}})$ is called the
coefficient-interval space or *parameter-interval space*. Thus to each
rounding IS_N a corresponding rounding IR_N is adjoined:

$$IS_N\colon \boldsymbol{P\mathcal{M}} \to IS_N(\boldsymbol{P\mathcal{M}}),$$
$$IR_N\colon \boldsymbol{P\mathcal{M}} \to IR_N(\boldsymbol{P\mathcal{M}}).$$

Analogous to (3.1-1), we introduce the following representation for intervals of functions $F \in IS_N(P\mathcal{M}) = II\, sp\{\phi_i\}_{i=0}^N \subset P\mathcal{M}$:

$$F = \Phi * A := \sum_{i=0}^N \phi_i A_i, \qquad (3.2\text{-}2)$$

where $A = (A_0, A_1, ..., A_N) \in IR_N(P\mathcal{M})$. $IR_N(P\mathcal{M})$ is some appropriate interval space, e.g., $II\mathbb{R}^N$, $II\,\mathbb{C}^N$,... . The analogue of (3.1-2) is the following correspondence for an $F \in IS_N(P\mathcal{M}) \subset P\mathcal{M}$:

$$\begin{aligned} IS_N(F) = IS_N(\Phi * A) &= \Phi * \mathcal{A}(IS_N) * A \\ &= \Phi * IR_N F, \end{aligned} \qquad (3.2\text{-}3)$$

where we have made the identification $IR_N F = \mathcal{A}(IS_N) * A$. $\mathcal{A}(IS_N)$ is a formal $\infty \times N'$ *interval matrix* (a matrix whose elements are intervals). In particular, $IS_N(\phi_j)$ is the jth column of $\mathcal{A}(IS_N)$, say $IS_N(\phi_j) = \Phi * IR_N(\phi_j)$. The comments in Section 3.1 concerning the range of S_N have their counterparts here. IS_N usually operates on a subset $IS_M(P\mathcal{M})$ of $P\mathcal{M}$ with $M \geq N$. Typically,

$$IS_N(P\mathcal{M}) \subset IS_M(P\mathcal{M}) \subset P\mathcal{M}. \qquad (3.2\text{-}4)$$

While M may be infinite, the range of IS_N is in every case of finite dimension. As an aid to conceptualization, view $IS_N(P\mathcal{M})$ as the space of finite *interval Fourier expansions,* that is, finite Fourier expansions with convex compact sets as coefficients. Likewise view $IR_N(P\mathcal{M})$ as $II\mathbb{R}^N$. Thus with $f \in \mathcal{M}$, $IS_N f$ in $IS_N(P\mathcal{M})$ is a finite interval Fourier expansion of f, and $IR_N f$ is the column vector in $II\mathbb{R}^{2N+1}$ that represents the coefficients of the Fourier expansion of $IS_N f$, viz.,

$$f \in IS_N(f) = \sum_{n=-N}^N C_n e^{inx}, \quad C_n \in II\mathbb{R},$$

$$IR_N f = C = (C_{-N}, ..., C_N)^T \in III\mathbb{R}^{2N+1}.$$

For example, suppose that $f = \sum_{-\infty}^{\infty} a_n e^{inx}$. Let $f_N = \sum_{n=-N}^N a_n e^{inx}$, and let $\bar{e}_N = \sup_{x \in X}(f - f_N)$ and $\underline{e}_N = \inf_{x \in X}(f - f_N)$. Set $C_0 = [a_0 - \underline{e}_N, a_0 + \bar{e}_N]$, $C_i = a_i, |i| > 1$. Then in the sense defined, we have $f \in \sum_{n=-N}^N C_n e^{inx}$.

This is not a particularly sharp inclusion, and in general, we would typically expect $f \in IS_N f \subset \sum_{n=-N}^{N} C_n e^{inx}$.

A schematic representation of these spaces and operators is given in Figs. 3.2-1 and 3.2-2.

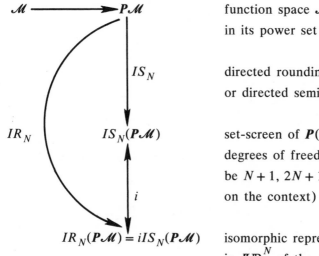

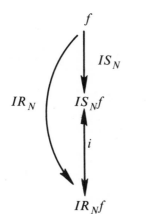

	function space $\mathcal{M}$ embedded in its power set
IS_N	directed rounding operator or directed semimorphism
$IS_N(P\mathcal{M})$	set-screen of $P(\mathcal{M})$ with N' degrees of freedom (N' might be $N+1$, $2N+1$, etc., depending on the context)
$IR_N(P\mathcal{M}) = iIS_N(P\mathcal{M})$	isomorphic representation in $I\!I\mathbb{R}^N$ of the screen.

Figure 3.2-1: Spaces, screens and mappings

f	function
$IS_N f$	directed round of the function f, i.e., an inclusion of f
$IR_N f$	isomorphic vector representation of the directed round of f

Figure 3.2-2: Directed round of f

We shall frequently write V for $IS_N f$ generically, and iV for $IR_N f$.

When no confusion will result we write V for iV.

From the given algebraic structure $(\mathcal{M}; +,-,\bullet,/,\int)$ we induce a corresponding structure $(P\mathcal{M}; +,-,\bullet,/,\int)$ in $P\mathcal{M}$. First we define operations in $P\mathcal{M}$.

For each $Y,Z \in P\mathcal{M}$

$$\bigwedge_{\circ \in \{+,-,\bullet,/,\int\}} Y \circ Z := \{y \circ z \mid y \in Y, z \in Z\}.$$

For the monadic operator $\int$, Y and y should be deleted here. A similar remark pertains to (3.2-5).

With this, we employ the *directed semimorphism* IS_N to induce an algebraic structure in $IS_N(P\mathcal{M})$, viz.,

$$\bigwedge_{\circ \in \{+,-,\bullet,/,\int\}} \bigwedge_{Y,Z \in IS_N(P\mathcal{M})} Y \diamondsuit Z := IS_N(Y \circ Z). \qquad (3.2\text{-}5)$$

(Recall that IS_N acting on a set in $P\mathcal{M}$ is defined simply by taking the union. We remark that $\diamondsuit$ has the property of isotoney: $Y \subset Z \Rightarrow \diamondsuit Y \subset \diamondsuit Z$ (cf. [11]). For computer realization of (3.2-5) see Section 7.1.4.) The induced structure is denoted by

$$(IS_N(P\mathcal{M}); \diamondsuit, \diamondsuit, \diamondsuit, \diamondsuit, \diamondsuit), \qquad (3.2\text{-}6)$$

and it is called an *(interval-)functoid*. Using the notation

$$IS_N(\Omega) = \{\diamondsuit, \diamondsuit, \diamondsuit, \diamondsuit, \diamondsuit\},$$

we may write the interval functoid (3.2-6) more simply as $(IS_N(P\mathcal{M}); IS_N(\Omega)) = IS_N(P\mathcal{M},\Omega)$ (compare (3.1-7)-(3.1-10)).

The isomorphism i maps the functoid into the interval space $IR_N(\mathcal{M})$. A substructure in this latter space, which we denote by

$$(IR_N(P\mathcal{M}); +,-,*),$$

is a well-known interval space (cf. [1], [11] where such structures are called vectoids of interval vectors). The operator $*$ denotes the inner product of two elements of the vectoid.

Thus for two sets $U, V \in P\mathcal{M}$ we define inclusion pointwise (cf. Section 2.2):

$$\bigwedge_{x \in X} U(x) \subset V(x) \Longleftrightarrow: U \subset V,$$

$$\bigwedge_{x \in X} U(x) \overset{\circ}{\subset} V(x) \Longleftrightarrow: U \overset{\circ}{\subset} V.$$

(Here we may exclude subsets of X of measure zero.)

If, in particular, U and $V \in IS_N(P\mathcal{M})$, then

$$U = \Phi * A \quad \text{and} \quad V = \Phi * B,$$

where $A = (A_0, ..., A_N)$, $B = (B_0, ..., B_N) \in IR_N(P\mathcal{M})$. Then the containment relations

$$U \subset V \text{ (resp. } U \overset{\circ}{\subset} V) \qquad\qquad (3.2\text{-}7)$$

are assured by the following conditions:

$$A \subset B \text{ (resp. } A \overset{\circ}{\subset} B), \qquad\qquad (3.2\text{-}8)$$

since

$$A \subset B \Rightarrow \Phi * A \subset \Phi * B$$

$$(\text{resp. } A \overset{\circ}{\subset} B \Rightarrow \Phi * A \overset{\circ}{\subset} \Phi * B).$$

(3.2-8) is easily implemented and tested in a computer.

In Table 3.2-1 we assemble and display a collection of spaces and mappings that constitute the setting of computer arithmetic. These include the spaces and roundings we have just introduced. They also include vector and matrix versions of these, which are appropriate for higher-dimensional problems. Finally, we find included the spaces in terms of which the first two classes are realized on a computer.

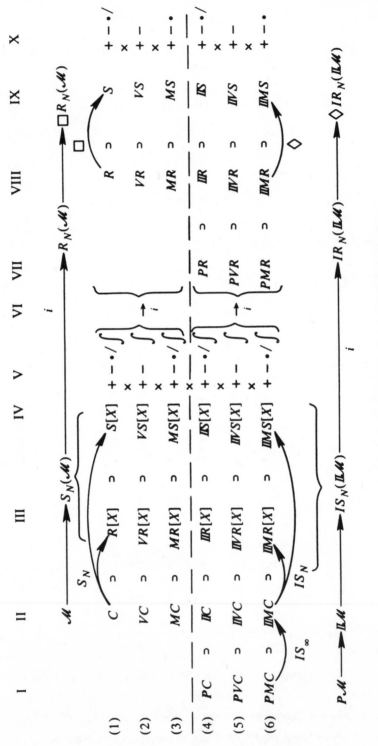

Table 3.2-1: Fundamental spaces of computation

We now explain the contents of Table 3.2-1 in some detail. There are six rows and ten columns in the table, the rows being divided into two principal classes. Rows (1), (2), and (3) comprise the standard data types, while rows (4), (5), and (6) comprise the interval data types. Rows (1) and (4) are the scalar types, rows (2) and (5) are the vector types (the prefix V or IV being employed), and rows (3) and (6) are the matrix types (the prefix M or IM being employed).

The following is a column-wise glossary for the table.

I   These are the power sets $P\mathcal{M}$ of the fundamental spaces in the first three rows of column II.

II   Rows (1), (2), and (3) contain the fundamental spaces $\mathcal{M}$ while rows (4), (5), and (6) contain special corresponding interval spaces $I\mathcal{M}$.

III, IV   These are screens $S_N(\mathcal{M})$ and $IS_N(I\mathcal{M})$ of the corresponding elements in column II, in particular, taken here to be polynomial screens and polynomial interval screens, as the case may be, each of degree not greater than N. In column IV the coefficients of the polynomials are, in particular, restricted to be computer representable.

V   These are the algebraic operations for the spaces of columns II, III, IV, rowwise. The operations, in particular, are to be interpreted semimorphically (cf. (3.1-8) and (3.2-5)). The symbol × between the operations in each row symbolizes the customary outer operations between spaces of different rows, e.g., vectors may be multiplied by scalars or by matrices, etc.

VI   Displays the isomorphism for computer representation of the spaces to the left (i.e., columns I-IV) by the spaces to the right (i.e., columns VII-IX).

VII, VIII   These are the coefficient spaces $R_N(\mathcal{M})$ (resp. $IR_N(I\mathcal{M})$). These constitute the collection of isomorphic images of the spaces listed in III. For example, VR is the isomorphic image of $VR[x]$.

IX   These are the computer-represented spaces $\Box R_N(\mathcal{M})$ and $\Diamond IR_N(I\!\!L\mathcal{M})$. They constitute the collection of isomorphic images of the spaces listed in column IV. We remark, in particular, that since column IV and column IX are isomorphic, we may hereafter regard the spaces in column IV as represented on a computer, i.e., as data types supplied with the associated algebraic operations as listed in column V.

X   These are the algebraic operations for the spaces of columns VII, VIII, IX, rowwise. The operations are to be interpreted semimorphically (see [11]). For the symbol × between rows, a description analogous to column V holds.

3.2.1 Examples of Bases

The bases listed in Section 3.1.1 may be used to generate bases for the interval-functoid. In particular, if $\{\phi_i\}_0^N$ is a given basis for a functoid, it is also a basis for an interval-functoid if there exists a directed rounding $IS_N\colon \phi_k \to I\!Isp\{\phi_i\}$, $N < k \leq M$, which we call the *inclusion property* of a basis Φ, i.e.,. $\phi_k \in IS_N\phi_k$ for all $k = 0, 1, \ldots N$.

In Section 3.2.2 we consider some such directed roundings.

3.2.2 Examples of Directed Roundings

As in Section 3.1.2 we are led to distinguish between the explicit and implicit cases.

(i)   EXPLICITLY DEFINED DIRECTED ROUNDINGS (problem independent)

We consider the situation in which the results of applying the operations in $IS_N(\Omega) = \{\diamondsuit, \diamondsuit, \diamondsuit, \diamondsuit, \oint\}$ to arguments taken from $IS_N(P\mathcal{M})$ are exactly representable in $IS_M(P\mathcal{M}) \supset IS_N(P\mathcal{M})$ for some $M \geq N$. According to (3.2-1), we must arrange that $\phi_k \in IS_N(\phi_k)$, $k = 0,\ldots,M$.

Now setting F in (3.2-3) to be ϕ_k, we get

$$IS_N(\phi_k) := \Phi * \mathscr{A}(IS_N) * i\phi_k, \quad k = 0,...,M.$$

Taking a canonical representation in $R_N(\mathscr{M})$ so that $i\phi_k = (\delta_{jk})$, the previous relation becomes

$$\Phi * \mathscr{A}(IS_N) * (\delta_{jk}) = \sum_{j=0}^{N} \phi_j A_{jk}, \quad k = 0,...,M, \qquad (3.2.2\text{-}1)$$

where A_{jk} is the kth column of $\mathscr{A}(IS_N)$ for all $0 \le k \le M$.

For simplicity, we typically choose a basis that is canonical with respect to the rounding IS_N, i.e., $IS_N(\phi_k) = \phi_k$, $k = 0,...,N$. In this basis the $N' \times M'$ interval matrix $\mathscr{A}(IS_N) := (A_{jk})$ is representable as

$$\mathscr{A}(IS_N) = (E, \overset{\wedge}{\mathscr{A}}(IS_N)),$$

where E is the $N' \times N'$ identity and $\overset{\wedge}{\mathscr{A}}(IS_N)$ is an $N' \times (M - N)$ interval matrix. If $M = N$, then $\overset{\wedge}{\mathscr{A}}(IS_N)$ is empty and $\mathscr{A}(IS_N) = E$. Referring to (3.2.2-1), we see that $\overset{\wedge}{\mathscr{A}}(IS_N)$ is composed of the last $M - N$ columns of (A_{jk}).

Methods of approximation theory supply estimates of the following sort:

$$\max_{x \in X} |\phi_k - S_N \phi_k| \le \sigma_k, \quad k = 0,...,M, \qquad (3.2.2\text{-}2a)$$

for many classes of bases (e.g., Chebyshev or Fourier bases). From such an estimate in turn we may specify a directed rounding as

$$IS_N(\phi_k) = S_N(\phi_k) + [-1, 1]\sigma_k, \quad k = 0,...,M. \qquad (3.2.2\text{-}2b)$$

In the canonical case $\sigma_k = 0$, $k = 0,...,N$.

For such a directed rounding, we can write $\mathscr{A}(IS_N)$ and $\overset{\wedge}{\mathscr{A}}(IS_N)$ in a more special form:

$$\mathscr{A}(IS_N) = (E, \overset{\wedge}{\mathscr{A}}(IS_N)) := (E, \overset{\wedge}{\mathscr{A}}(S_N)) + (\mathscr{O}, [-1, 1]D_N).$$

Here $\hat{A}(S_N)$, D_N, and $\mathcal{O}$ are customary point matrices (i.e., not interval matrices). In particular, D_N is an $N' \times (M - N)$ matrix and $\mathcal{O}$ is a block of zeros. Referring to (3.2.2-2), we see that the kth column of D_N is $(\sigma_{N+k}, 0, 0, \ldots)^T$.

Let us now consider some particular roundings.

Directed Taylor Rounding

In the *MPB* case we may specify the directed rounding IS_N as follows:

$IS_N(x^{N+j}) := S_N(x^{N+j}) + [-1, 1]\sigma_j$, $j \geq 1$ (cf. section 3.1.2). Indeed for the domain $[-1, 1]$, all $\sigma_j = 1$, and we take $\hat{\mathcal{A}}(S_N) = 0$ (zero matrix) and

$$
D_N = \begin{pmatrix}
1 & . & . & . & 1 \\
0 & . & . & . & 0 \\
. & . & . & . & . \\
. & . & . & . & . \\
. & . & . & . & . \\
0 & . & . & . & 0
\end{pmatrix}. \qquad (3.2.2\text{-}3)
$$

Directed Chebyshev Rounding

Table 3.1.2-1 lists the quantities σ_j needed to specify the Chebyshev directed rounding. For example, $\mathcal{A}(S_4)$ and D_4 needed to define IS_4 may be composed from this table and are, in particular:

$$
\hat{\mathcal{A}}(S_4) := \begin{pmatrix}
0 & 1/32 & 0 & 7/128 & \cdots \\
-5/16 & 0 & -28/64 & 0 & \cdots \\
0 & -18/32 & 0 & -112/128 & \cdots \\
20/16 & 0 & 84/64 & 0 & \cdots \\
0 & 48/32 & 0 & 224/128 & \cdots \\
0 & 0 & 0 & 0 & \cdots \\
. & . & . & . & \cdots \\
. & . & . & . & \cdots \\
. & . & . & . & \cdots
\end{pmatrix},
$$

$$D_4 := \begin{pmatrix} 1/16 & 1/32 & 8/64 & 9/128 & \cdots \\ 0 & 0 & 0 & 0 & \cdots \\ \cdot & \cdot & \cdot & \cdot & \cdots \\ \cdot & \cdot & \cdot & \cdot & \cdots \\ \cdot & \cdot & \cdot & \cdot & \cdots \end{pmatrix},$$

where the jth column of $\hat{\mathcal{A}}(S_4)$ is $R_4(x^{j+4}) = iS_4(x^{j+4})$, $j = 1,2,\ldots$. For example, the first column of $\hat{A}(S_4)$ corresponds to $iS_4(x^5)$. Referring to Table 3.1.2-1, we find the polynomial $20x^3 - 5x$ in the entry corresponding to $(16x^5, S_4(\phi_j))$. The first column vector comes from this polynomial divided by 16, i.e., $0, -5, 0, 20, 0, \ldots)^T/16$. As we have observed earlier, the kth column of D_4 is $(\sigma_{4+k},0,0,\ldots)^T$. Referring to Table 3.1.2-1 again we see, for instance, that σ_{4+1} is found in the σ_j entry corresponding to $(16x^5, S_4(\phi_j))$ and is, in particular, 1 (scaled by $1/16$).

Directed Spline Rounding

The case of directed spline roundings is constructed from an appropriate composition of the spline roundings. The latter were discussed above in Section 3.1.2 to which we now refer.

For all $P\mathcal{M}$ we consider the representation

$$Y(x) = \sum_{i=1}^{N} A_i(x)\mathcal{X}_i, \quad A = (A_1,\ldots,A_N) \in IR_N(P\mathcal{M}). \quad (3.2.2-5)$$

$IR_N(P\mathcal{M})$ will typically be $P(R_N(\mathcal{M}))$. Hence with $R_N(\mathcal{M}) = \mathcal{N}$ we have $IR_N(P\mathcal{M}) = (P\mathcal{N})^N$ or $IR_N(P\mathcal{M}) = (\mathbb{I}\mathcal{N})^N$. Here in analogy to Section 3.1.2 we could replace $A \in IR_N(P\mathcal{M})$ by an interval $A \in (IR_N(P\mathcal{M}), \mathcal{BC})$ to denote conditions imposed on the coefficients A_i. We shall presently make use of another such function $Z \in \mathcal{X}^{\mathbb{I}R_N(P\mathcal{M})}$. $B_i(x)$, $i = 1,\ldots,N$, will denote the coefficients of Z.

Corresponding to the structure $(R_N(\mathcal{M}); +,-,\bullet,/,\int)$, a structure $(IR_N(P\mathcal{M}); +,-,\bullet,/,\int)$ is induced in the power set $IR_N(P\mathcal{M})$. Similarly a corresponding structure $(\mathscr{X}^{IIR_N(P\mathscr{M})}; +, -, \bullet, /, \int)$ is induced in $\mathscr{X}^{IIR_N(P\mathscr{M})}$. In particular, for the latter structure, we have

$$\bigwedge_{\substack{\circ \in \{+, -, \bullet, /\}}} \bigwedge_{Y, Z \in \mathscr{X}^{IIR_N(P\mathscr{M})}} Y \circ Z = \sum_{i=1}^{N} (A_i \circ B_i)\mathscr{X}_i, \qquad (3.2.2\text{-}6)$$

while

$$\int_{\xi}^{x} Y = \sum C_i \mathscr{X}_i. \qquad (3.2.2\text{-}7)$$

where, corresponding to the interval of $X = [a,b]$ and its partition as described following (3.1.2-7) in Section 3.1.2, we have $\xi := x_{k+1}$ and

$$C_i(x) \quad := C_{i-1}(x_i) + \int_{x_i}^{x} A_i(x)dx, \qquad x \in X_i, \ i \geq k+2,$$

$$C_{i+1}(x) := \int_{\xi}^{x} A_{k+1}(x)dx, \qquad x \in X_{k+1}, \qquad (3.2.2\text{-}8)$$

$$C_i(x) \quad := C_{i+1}(x_{i+1}) + \int_{x_{i+1}}^{x} A_i(x)dx, \quad x \in X_i, \ i \leq k.$$

We remark once again that the restriction made here to intervals is one of convenience only, and that any other relevant set and partition may be used.

Set inclusion in $IR_N(P\mathscr{M})$: $(IR_N(P\mathscr{M}), \subset)$ induces a corresponding ordering in $\mathscr{X}^{IR_N(P\mathscr{M})}$: $(\mathscr{X}^{IR_N(P\mathscr{M})}, \subset)$ as

$$\bigwedge_{Y, Z \in \mathscr{X}^{P\mathscr{X}}} Y \subset Z :\Longleftrightarrow \bigwedge_{1 \leq i \leq N} A_i \subset B_i. \qquad (3.2.2\text{-}9)$$

Suppose corresponding to (3.1.2-12) that the rounding IS_N is defined coefficientwise in terms of the directed rounding $IS: P\mathscr{N} \rightarrow IS(P\mathscr{N})$, viz.,

$$IS_N(\sum_{i=1}^{N} A_i \mathscr{X}_i) = \sum_{i=1}^{N} IS(A_i)\mathscr{X}_i. \qquad (3.2.2\text{-}10)$$

Then we see that correspondingly $IR_N(P\mathscr{M}) = (IS(\mathscr{N}), ..., IS(\mathscr{N})) = (IS(\mathscr{N}))^{N}.$

Continuing, we have the following analogue of (3.1.2-13):

$$\mathcal{X}^{\text{IIR}}{}_N(\boldsymbol{P\mathcal{M}}) = \left\{ \sum_{i=1}^{N} A_i \mathcal{X}_i \mid A_i \in \boldsymbol{P\mathcal{N}}_i \right\} \qquad (3.2.2\text{-}11)$$

The analogue of (3.1.2-15) is

$$IS_N(\sum_{i=1}^{N} A_i \mathcal{X}_i) := \sum_{i=1}^{N} IS_{N_i,i}(A_i)\mathcal{X}_i, \qquad (3.2.2\text{-}12)$$

where

$$IS_{N_i,i}: \boldsymbol{P\mathcal{N}}_i \rightarrow IS_{N,i}(\boldsymbol{P\mathcal{N}}_i), \qquad (3.2.2\text{-}13)$$

and the $IS_{N,i}(\boldsymbol{P\mathcal{M}})$ are set screens of $\boldsymbol{P\mathcal{M}}$, which may be selected independently. They may, for instance, be chosen from among the examples already discussed.

example

An example of a directed spline rounding is composed from the one in Section 3.1.2 (cf. (3.1.2-15)f). $IS_{2,1}$ is defined by employing $X_1 = [0,1]$ and a directed Chebyshev rounding, while $IS_{2,2}$ is defined by employing $X_2 = [1,\infty)$ and an appropriately transformed Chebyshev rounding.

(ii) IMPLICITLY DEFINED DIRECTED ROUNDINGS (problem dependent)

In principle every implicit rounding, such as those introduced in Section 3.1.2(ii), can be used to construct an implicit directed rounding. However, some pragmatic difficulties concerning the quality of the inclusion (cf. (3.1.2-29)f) may arise. For example, following (3.2.2-2a), we compute the following error estimates that characterize this quality:

$$\sigma_0 = \| 3 - \frac{1}{3}x + \frac{2}{3}x^2 \|_\infty = 4,$$
$$\sigma_1 = \| \frac{2}{3} - \frac{5}{3}x + \frac{2}{15}x^2 \|_\infty = \frac{37}{15}. \qquad (3.2.2\text{-}14)$$

Similarly, we find $\sigma_2 = 34/15$ and $\sigma_3 = 44/21$.

We defer treatment of this particular difficulty to future work, and we confine our discussion to a single case: the optimal implicitly defined directed rounding.

Let $p(\mathscr{L},V,IS_N)$ be a process that employs the rounding IS_N and that delivers an inclusion $V(x)$ of the solution y of $\mathscr{L}y = 0$. (Such processes will be dealt with in more detail in Chapter 4.)

We require IS_N to be optimal in a certain sense, e.g., letting $d(V(x))$ denote the diameter of the set $V(x)$ at the point x,

$$\max_{x \in X} d(V(x)) = \| d(V(x)) \|_\infty = \min, \qquad (3.2.2\text{-}15)$$

within the context and setting of IS_N (such as N, choice of basis, coefficient space). Expressing V as a function of p, $\mathscr{L}$ and IS_N, (3.2.2-15) becomes more formally

$$\| d(V(p,\mathscr{L},IS_N,x)) \| = \min . \qquad (3.2.2\text{-}16)$$

Evidently this is an implicit nonlinear definition of IS_N as an optimal directed rounding.

(iii) DIRECTED ROUNDING WITH CONSTRAINTS

Constraints of the form $c(y) = 0$ for $y \in \mathcal{M}$ as given in Section 3.1.2(iii) induce corresponding constraints in the power set $P\mathcal{M}$. Corresponding to a $Y \in P\mathcal{M}$ we distinguish between the following two principle types of such power set constraints (a) and (b):

(a) $\qquad \bigwedge_{y \in Y} c(y) = 0,$

which is abbreviated as

$$c(Y) = 0. \qquad (3.2.2\text{-}17)$$

(b) Suppose c depends on the parameters $a = (a_1,...,a_N)$ with $a_i \in \mathbb{R}$. Then corresponding to the constraint $c(a,y) = 0$ imposed on the function $y \in \mathcal{M}$ for each $a_i \in A_i \in P\mathbb{R}$, we define (with $A = (A_1,...,A_n)$)

$$\bigwedge_{a \in A} \bigvee_{y \in Y} c(a,y) = 0.$$

This last is abbreviated by

$$\{c(a, y) = 0 \,|\, a \in A, y \in Y\}. \tag{3.2.2-18}$$

An example for (3.2.2-17) is

$$Y(0) = Y(1) = 0.$$

In this case both $Y(0)$ and $Y(1)$ must be point intervals, say $Y(0), Y(1) \in \mathbb{R}$.

An example for (3.2.2-18) is

$$A * Y(0) - B * Y(1) - C = 0.$$

This equation is to be interpreted a set of constraints to be fulfilled by $Y(0)$ and $Y(1)$:

$$\{a \cdot u - b \cdot v - c = 0 \,|\, a \in A, b \in B, c \in C \text{ and } u \in Y(0), v \in Y(1)\}.$$

Continuing to follow the development of Section 3.2.1(iii), we say that a directed rounding $IS_N: P\mathcal{M} \to IS_N(\mathbb{R}\mathcal{M})$ fulfills a constraint $c = 0$ (corresponding respectively to the constraints (a) and (b) here, i.e., to (3.2.-17) and (3.2.2-18)) if

$$\bigwedge_{Y \in P\mathcal{M}} c(IS_N(Y)) = 0 \tag{3.2.2-19}$$

resp.

$$\bigwedge_{Y \in P\mathcal{M}} \{c(a,z) = 0 \,|\, a \in A, z \in IS_N(Y)\}. \tag{3.2.2-20}$$

The methodology for constructing a directed rounding that satisfies such constraints is analogous to that given in Section 3.1.2(iii) replacing $\tilde{S}_N$ and S_N used there by $\tilde{IS}_N$ and IS_N here.

examples

1. As an example corresponding to the *PMB* basis $\{1,x,x^2,...\}$ and the constraint

$$c(Y) :\equiv Y(0) = 0,$$

we have the rounding

$$IS_N Y := x\widetilde{IS}_{N-1} Y.$$

2. In the same PMB basis consider the constraint

$$c(Y) = (Y(0) - \eta,\ Y(1) - \theta)^T.$$

The resulting rounding is

$$IS_N Y := \eta(1 - x) + \theta x + x(1 - x)\widetilde{IS}_{N-2} Y.$$

3. In the EMB basis consider the following constraint:

$$c(Y) = (Y(\infty),\ \int_0^\infty Y dt - 1)^T.$$

The resulting rounding is

$$IS_N Y := e^{-x}\widetilde{IS}_{N-1} Y - N\left(\int_0^\infty e^{-t}\widetilde{IS}_{N-2} Y dt - 1 \right)e^{-Nx}.$$

(iv) DIRECTED RELATIVE ROUNDINGS

The concepts, terminology, and usage for directed relative and nonrelative roundings are analogous to those given in Section 3.2.1(i). The terminology and concepts of mantissa, shifting factor, and the like are the same here. The rounding S_N^p is replaced by the directed relative rounding IS_N^p, and so forth. In particular if

$$Y = \sum_{i=0}^M A_i \phi_i,$$

then

$$IS_N^p Y = \phi_p S_N \sum_{j=0}^{M-p} a_{j+p} \phi_j \qquad (3.2.2\text{-}21)$$

(compare (3.1.2-43)).

Chapter 4

METHODS FOR FUNCTIONAL EQUATIONS

We begin Chapter 4 with the discussion of a model problem in order to illustrate our point of view and expose some of its methodology. Discussion of the critical and detailed steps of solution validation is deferred to the body of the chapter itself.

Let $\mathcal{M}$ be one of the spaces introduced in Chapter 3, and let $\mathcal{L}: \mathcal{M} \to \mathcal{M}$ be an operator. We seek a solution of the functional equation

$$\mathcal{L}y \equiv 0. \tag{4-1}$$

(Note that when it is necessary to distinguish between functional equality in $\mathcal{M}$ and pointwise equality we use $\equiv$ for the former and $=$ for the latter.)

As it stands (4-1) is not generally amenable to approximation computationally. For this purpose, we replace (4-1) by the modified problem

$$S_N \mathcal{L} v \equiv 0. \tag{4-2}$$

S_N is chosen so that (4-2) has a solution in $S_N(\mathcal{M})$ that is readily obtainable in a computational sense. (Compare this procedure with the notion of a weak solution or that of a Galerkin approximation (cf. (3.1.2-16)f).) Let us interpret (4-2) as a problem given in terms of a modified operator $\mathcal{L}_N := S_N \mathcal{L}$, so that (4-2) becomes

$$(S_N \mathcal{L}) v \equiv \mathcal{L}_N v \equiv 0. \tag{4-3}$$

For practical reasons, we sometimes prefer to view (4-3) as

$$S_N(\mathcal{L} v) \equiv 0. \tag{4-4}$$

Three questions arise:

(i) How can we demonstrate the existence of the solution y and estimate of the error $e_N := y - v$?

(ii) How are the operator $\mathcal{E}_N := \mathcal{L} - \mathcal{L}_N$ and the error e_N related?

(iii) How can we compute e_N, a posteriori, from the problem data (i.e., from S_N and $\mathcal{L}$) and the approximation v?

We take the following model problem to illustrate realization of (4-3) and (4-4) as well as other points introduced here:

$$\mathcal{L} y \equiv y - \int_0^x y \, dx - 1 \equiv 0. \tag{4-5}$$

For clarity, we avoid using the self-validation methodology for demonstrating existence and uniqueness of a solution of (4-5) since this simple model problem is known to have the unique solution $y(x) = e^x$ in $\mathcal{M} = C^\infty$. For convenience, we confine our attention to the domain $X = [-1, 1]$. We set $N = 2$ and take $S_2(\mathcal{M}) = sp\{1, x, x^2\}$. The solution in $S_2(\mathcal{M})$ that we seek has the form

$$v = a + bx + cx^2$$

with the scalars a, b, and c to be determined. Correspondingly,

$$iv = (a,b,c)^T$$

in $R_N(\mathcal{M}) = R_2(\mathcal{M}) = \mathbb{R}^3$.

According to (4-4), v must be a solution of

$$S_2(v - \int_0^x vdx - 1) \equiv 0.$$

Thus

$$S_2(h(x) - \frac{1}{3}cx^3) \equiv 0, \tag{4-6}$$

where

$$h(x) = a - 1 + (b - a)x + (c - \frac{1}{2}b)x^2.$$

According to the property (3.1-0), $S_2h(x) = h(x)$. Then (4-6) becomes

$$h - \frac{1}{3}cS_2(x^3) = 0. \tag{4-7}$$

Let us employ the Chebyshev rounding introduced in Chapter 3. Then $S_2(x^3) \equiv \frac{3}{4}x$, and so, (4-7) becomes

$$a - 1 + (b - a - \frac{1}{4}c)x + (c - \frac{1}{2}b)x^2 \equiv 0.$$

This implies that

$$a \qquad\qquad = 1,$$

$$-a + b - \frac{1}{4}c = 0,$$

$$-\frac{1}{2}b + c = 0,$$

so that $iv = \left(1, \frac{8}{7}, \frac{4}{7}\right)^T$. Thus

$$v(x) = 1 + \frac{8}{7}x + \frac{4}{7}x^2. \tag{4-8}$$

As determined, $v(x)$ approximates e^x with an error

$$|e_2| \le 0.07. \tag{4-9}$$

The following is a table of $e_2(x)$.

x	$e_2(x)$
1.00	0.004
0.75	-0.037
0.60	-0.070
0.50	-0.066
0.25	-0.062
0.00	± 0.000
-0.50	0.036
-1.00	-0.061

Since $\mathscr{L}y = \mathscr{L}_N v = 0$, then $\mathscr{E}_N v \equiv (\mathscr{L}_N - \mathscr{L})v \equiv \mathscr{L}y - \mathscr{L}v \equiv \mathscr{L}e_N$. From this we may write, in particular, that

$$e_2 \equiv \mathscr{L}^{-1}\mathscr{E}_2 v_2, \tag{4-10}$$

where the subscript on v here indicates that in this relation v is particularized to the case $N = 2$.

Using (4-5) and (4-8), we may show that

$$\mathscr{L}v_2 \equiv -\frac{1}{21}T_3(x),$$

and since $\mathscr{L}_2 v_2 \equiv 0$, we also have that

$$(\mathscr{L}_2 - \mathscr{L})v_2 \equiv \mathscr{E}_2 v_2 \equiv -\frac{1}{21}T_3(x).$$

Then from (4-9) and (4-10)

$$\|y - v_2\| \le \frac{1}{21}\|\mathscr{L}^{-1}\|\,\|T_3\|.$$

Based on this example, we make several observations, the first of which is a key remark concerning our computational point of view, i.e., validation.

Remarks:

(i)   This example illustrates a customary viewpoint of error analysis which is to estimate the error (viz., $\|e_2\|$) in terms of both the residual (viz., $\mathscr{L}v_2 \equiv - \frac{1}{21}T_3(x)$) and $\|\mathscr{L}^{-1}\|$. As we shall presently see, our approach is to obtain an error estimate a posteriori as a by-product of the computation itself. It is the validation process itself that provides the a posteriori estimate. In this way we to avoid determining the a priori estimate of $\|\mathscr{L}^{-1}\|$, which is typically no less difficult to obtain than the solution itself. Moreover, even when it is obtainable the a priori estimate is typically so pessimistic as to be of little practical value.

(ii)   The ansatz $v \in S_N(\mathscr{M})$ implies that (4-2) is solvable for such a v. Since $S_N(\mathscr{M})$ is a manifold described by N' degrees of freedom, v must be supplied with no fewer degrees of freedom.

A more general approach than the one that we have given for deriving a linear system for v will be the subject of the following Section 4.1. In Section 4.2 an alternative such derivation for the nonlinear case will be given, employing analogous principles.

(iii)   Our methodology requires determining the round of the residue

$$w \equiv \mathscr{L}v \in \mathscr{M}.$$

w itself is not typically an element of the screen $S_N(\mathscr{M})$ (for any particular value of N). A convenient class of problems to which we shall at first restrict ourselves consists of those problems for which the screen can be chosen so that

$$\mathscr{L}v \in S_M(\mathscr{M})$$

for some $M \in \mathbb{N}, M \geq N$. We then seek an appropriate modified operator $\widetilde{\mathscr{L}}$ or equivalently an intermediate representation $\widetilde{w}$ of the function w (such that $\mathscr{L}v = \widetilde{w}$) with the property

$$S_N\widetilde{w} \equiv S_N\widetilde{\mathscr{L}}v \equiv S_N\mathscr{L}v \equiv S_Nw.$$

(This procedure generalizes to our setting for ultra-arithmetic, an approach to computer arithmetic and the rounding of numbers introduced in [12].) For example, let $\mathcal{L}^*$ satisfy the condition

$$\mathcal{L}^* v \in \mathcal{M} \setminus S_N(\mathcal{M}).$$

Then by employing an appropriate $\mathcal{L}^*$, $\widetilde{\mathcal{L}}$ may be constructed in the form

$$\widetilde{\mathcal{L}} := \mathcal{L} - \mathcal{L}^*.$$

(iv) In the balance of this monograph we restrict ourselves to problems of algebraic type. This means that a generic problem

$$\mathcal{L} y \equiv 0, \qquad\qquad\qquad (4\text{-}11)$$

for $y \in \mathcal{M} = C_x^0$ say, is expressed algebraically in terms of the operations in $(\mathcal{M}; +, -, \bullet, /, \int)$. We emphasize that this restriction does not confine us to algebraic operators. Indeed, transcendental operators of many sorts are captured by this framework. For details, see the introductory comments of Section 4.2.2.

Having completed our preliminary discussion, we shall continue to develop the techniques and tools of our methodology, usually in the context of typical model problems. First, we treat the linear case in Section 4.1, and then in Section 4.2 we treat the nonlinear case. In both cases approximation in the standard sense is given first followed by the validation step, that is, by computation employing directed roundings and supplying inclusions.

4.1 METHODS FOR LINEAR EQUATIONS

Let $\mathcal{M}$ be a linear space and Φ be a basis. For each $w \in \mathcal{M}$, we have

$$w = \Phi * iw$$

(cf. (3.1-1)). Corresponding to a linear operator $\mathcal{L}: \mathcal{M} \to \mathcal{M}$, we have

$$\mathcal{L}(w) := \mathcal{L}(\Phi * iw) := \Phi * \mathcal{A}(\mathcal{L}) * iw.$$

We suppose that the basis Φ is chosen so that for every $v \in S_N(\mathcal{M})$, there exists an $M \in \mathbf{N}$ with $\mathcal{L}v \in S_M(\mathcal{M})$. We call the difference $\delta = M - N$ the *inflation index* of $\mathcal{L}$. We shall consider only operators $\mathcal{L}$ for which $\delta \geq 0$. In a sense, these operators correspond to the well-posed problems.

Recalling the isomorphism $i: S_N(\mathcal{M}) \longleftrightarrow R_N(\mathcal{M})$, we see that the expression

$$i(S_N\mathcal{L}): S_N(\mathcal{M}) \underset{\mathcal{L}}{\to} S_M(\mathcal{M}) \underset{S_N}{\to} S_N(\mathcal{M}) \underset{i}{\to} R_N(\mathcal{M})$$

may be viewed as a mapping of $S_N(\mathcal{M}) \to R_N(\mathcal{M})$. A fortiori $iS_N\mathcal{L}i^{-1}: R_N(\mathcal{M}) \to R_N(\mathcal{M})$. Such mappings are represented by matrices. Then by means of

$$i(S_N\mathcal{L}v) := \mathcal{A}(S_N)\mathcal{A}(\mathcal{L})*iv, \tag{4.1-1}$$

from which we deduce

$$iS_N\mathcal{L}i^{-1} = \mathcal{A}(S_N)\mathcal{A}(\mathcal{L}),$$

the matrix $\mathcal{A}(\mathcal{L})$ is defined. Generally we may write

$$i\mathcal{K}i^{-1} =: \mathcal{A}(\mathcal{K}) \tag{4.1-2}$$

for $\mathcal{K}$ an operator (e.g., $\mathcal{K} = \mathcal{L}, \ell, S_N,...$) (cf. (3.1.2-2)f.).

Remark: Refer to the remark following (3.1-2). Here we note that $\mathcal{A}(\mathcal{L})$ is a formal $\infty \times \infty$ matrix. In practice $\mathcal{A}(\mathcal{L})$ will be applied to N-dimensional vectors $iv \in iS_N(\mathcal{M})$ yielding M-dimensional vectors (possibly infinite) in $iS_M(\mathcal{M})$. In practice it suffices to take the corresponding $M' \times N'$ block of $\mathcal{A}(\mathcal{L})$, neglecting thereby all formally zero terms. We shall hereafter do this without special notation for $\mathcal{A}(\mathcal{L})$. The context should eliminate any confusion.

$\mathcal{A}(S_N)$ is an $M' \times N'$ matrix taking $R_M(\mathcal{M})$ into $R_N(\mathcal{M})$. For convenience, we identify $\mathcal{A}(S_N)$ with R_N. According to (3.1-0), the rounding operator S_N is the identity operator on $S_N(\mathcal{M})$. Thus the matrix

$\mathcal{A}(S_N)$ has the blockwise form

$$\mathcal{A}(S_N) = (E, \hat{\mathcal{A}}(S_N))$$

(compare (3.1.1-2)). E is the $N' \times N'$ identity matrix, while $\hat{\mathcal{A}}(S_N)$ is the $N' \times \delta$ matrix that actually implements the rounding operation. Of course, if $\delta = 0$, then $\hat{\mathcal{A}}(S_N)$ is empty.

Similarly $\mathcal{A}(\mathcal{L})$ may be decomposed blockwise as follows:

$$\mathcal{A}(\mathcal{L}) = \begin{pmatrix} \overset{\circ}{\mathcal{A}}(\mathcal{L}) \\ \hat{\mathcal{A}}(\mathcal{L}) \end{pmatrix}. \tag{4.1-3}$$

$\overset{\circ}{A}(\mathcal{L})$ is an $N' \times N'$ matrix, while $\hat{A}(\mathcal{L})$ is an $\delta \times N'$ matrix. Then

$$\mathcal{A}(S_{\overline{N}})\mathcal{A}(\mathcal{L}) = E\overset{\circ}{\mathcal{A}}(\mathcal{L}) + \hat{\mathcal{A}}(S_N)\hat{\mathcal{A}}(\mathcal{L}) \tag{4.1-4}$$

is an $N' \times N'$ matrix. The decomposition displayed in (4.1-4) plays a key part in automatic error control. It also serves as a guide for theoretical study of decomposition techniques and of rounding operators in the linear setting as well as a guide for the nonlinear case.

We are now concerned with explicit determination of the matrix $\mathcal{A}(\mathcal{L})$. We distinguish the finite case in Section 4.1.1 and the infinitesimal case in Section 4.2.2. The latter case covers differentiation, integration, shift operators, and the like, whereas the former excludes such operations.

4.1.1 The Finite Case

Let B be a function space, and let $\mathcal{M} = B^n$ be the n-fold direct product. We seek a solution $y(x) = (y_1(x),...,y_n(x))^T$ in $\mathcal{M}$ of the linear problem

$$\mathcal{L}(x)y(x) \equiv \ell(x)y(x) - b(x) \equiv 0.$$

$\ell(x)$ and $b(x)$ are given, and in particular, $b(x) = (b_1(x),...,b_n(x))^T$ and $\ell(x) = (\ell_{ij}(x))$. Here $\ell: \mathcal{M} \to \mathcal{M}$ represents an $n \times n$ function-valued matrix.

We continue with the case $n = 2$, which is adequate for the exposition. The rounded problem (cf. (4-4))

$$S_N(\ell(x)v(x) - b(x)) = 0$$

is then simply

$$S_N\left\{\begin{pmatrix} \ell_{11}(x) & \ell_{12}(x) \\ \ell_{21}(x) & \ell_{22}(x) \end{pmatrix} * \begin{pmatrix} v_1(x) \\ v_2(x) \end{pmatrix} - \begin{pmatrix} b_1(x) \\ b_2(x) \end{pmatrix}\right\} = 0. \quad (4.1.1\text{-}1)$$

The $*$ is used to denote matrix-vector product here and in what follows. This is technically a different use of the symbol $*$ which is employed in Section 3.1.

The linear rounding operator may be taken to be a 2×2 matrix of rounding operators,

$$S_N = \begin{pmatrix} S_{N,11} & S_{N,12} \\ S_{N,21} & S_{N,22} \end{pmatrix},$$

where

$$S_{N,ij} : B \rightarrow S_{N,ij}(B), \quad 1 \le i,j \le 2.$$

In practice, all $S_{N,ij} = 0$ if $i \neq j$ and all $S_{N,ii}$ are equal to the same operator which we denote also by S_N for convenience. Thus (4.1.1-1) becomes

$$S_N(\ell_{11}(x)v_1(x) + \ell_{12}(x)v_2(x) - b_1(x)) = 0,$$

$$S_N(\ell_{21}(x)v_1(x) + \ell_{22}(x)v_2(x) - b_2(x)) = 0.$$

Upon employing the isomorphism $i: S_N(B) \rightarrow R_N(B)$ these equations become (the analogue of (4.1-1)) the following linear system in $R_N(B) \times R_N(B)$:

$$\begin{pmatrix} \mathcal{A}(S_N) & 0 \\ 0 & \mathcal{A}(S_N) \end{pmatrix}$$

$$(4.1.1\text{-}2)$$

$$\times \left\{ \begin{pmatrix} \mathcal{A}(\ell_{11}) & \mathcal{A}(\ell_{12}) \\ \mathcal{A}(\ell_{21}) & \mathcal{A}(\ell_{22}) \end{pmatrix} * \begin{pmatrix} iv_1 \\ iv_2 \end{pmatrix} - \begin{pmatrix} ib_1 \\ ib_2 \end{pmatrix} \right\} = \begin{pmatrix} 0 \\ 0 \end{pmatrix}.$$

In the general case, we may rewrite (4.1.1-1)

$$\begin{pmatrix} \mathcal{A}(S_{N,11})\mathcal{A}(\ell_{11}) & \mathcal{A}(S_{N,11})\mathcal{A}(\ell_{12}) \\ \mathcal{A}(S_{N,22})\mathcal{A}(\ell_{21}) & \mathcal{A}(S_{N,22})\mathcal{A}(\ell_{22}) \end{pmatrix} * \begin{pmatrix} iv_1 \\ iv_2 \end{pmatrix} = \begin{pmatrix} \mathcal{A}(S_{N,11})ib_1 \\ \mathcal{A}(S_{N,22})ib_2 \end{pmatrix}.$$

From this equation it is clear that the n-dimensional case is a straight-forward generalization of the one-dimensional case treated earlier.

example

We continue further by specializing to the following example:

$$\begin{pmatrix} 1 & 2x \\ 1 & x \end{pmatrix} \begin{pmatrix} y_1 \\ y_2 \end{pmatrix} = \begin{pmatrix} 1 \\ 0 \end{pmatrix}, \qquad (4.1.1\text{-}3)$$

whose exact solution is

$$\begin{pmatrix} y_1 \\ y_2 \end{pmatrix} = \begin{pmatrix} -1 \\ 1/x \end{pmatrix}.$$

Let us seek an approximation on the screen $sp\Phi_N$ with $\Phi_N = \{1, x, x^2\}$ (so that $N = 2$) and on the domain $X = [2,4]$. Let

$$(iv_1 =)v_1 = a + bx + cx^2 \text{ and } (iv_2 =)v_2 = d + ex + fx^2.$$

We proceed to the system

$$S_2\left(a + (b + 2d)x + (c + 2e)x^2 + 2fx^3\right) = 1,$$

$$(4.1.1\text{-}4)$$

$$S_2\left(a + (b + d)x + (c + e)x^2 + fx^3\right) = 0.$$

We use the Chebyshev rounding, which on $X = [2,4]$ gives

$$S_2(x^3) = 9x^2 - 26.25x + 24.75. \qquad (4.1.1\text{-}5)$$

With this rounding, (4.1.1-3) becomes the following linear system given in tabular form:

a	b	c	d	e	f	rhs
1	0	0	0	0	49.50	1
0	1	0	2	0	-52.50	0
0	0	1	0	2	18.00	0
1	0	0	0	0	24.75	0
0	1	0	1	0	26.25	0
0	0	1	0	1	9.00	0

The solution of this linear system is

$$(a, b, c, d, e, f) = (-1, 0, 0, \frac{26.25}{24.75}, -\frac{9}{24.75}, \frac{1}{24.75}).$$

Thus

$$v_1 = -1,$$
$$v_2 = \frac{26.75 - 9x + x^2}{24.75}. \tag{4.1.1-6}$$

The error $e_1 = y_1 = 0$, while for the error $e_2 = y_2 - v_2$, we have $|e_2| \leq 0.005$. The following table of values is instructive:

x	v_2	e_2 (approx)
2	12.15/24.75	-0.005
3	8.25/24.74	0
4	6.25/24.75	0.0025

$$(4.1.1-7)$$

Such a detailed treatment of the approximate solution $(v_1, v_2)^T$ and its associated error is generally unavailable. Thus we are led to consider *automatic computer determination of errors,* that is to say, self-validation. We note in passing that $(v_1, v_2)^T$ is a functional approximation of the first column of the inverse of the matrix in (4.1.1-3).

VALIDATION (automatic error control)

We recall some notation introduced in Chapter 3. IS_N is the interval rounding operator delivering inclusions for the object rounded (cf.

(3.2-1)), viz.,

$$\bigwedge_{y \in \mathcal{M}} y \in IS_N y.$$

Actually the use required of (3.2-1) now is more particularly

$$\bigwedge_{w \in S_M(\mathcal{M})} w \in IS_N w.$$

We now seek an iteration method as described in Section 2.2. Let $\tilde{y} \in S_N(\mathcal{M})$ be an approximation to the desired solution y of the equation $\mathcal{L}y \equiv 0$. ($\tilde{y}$ may be calculated, guessed at, or obtained in any other way.) We next calculate the correction Y, an interval, so that $y \in \tilde{y} + Y$. Y is calculated by means of the following recurrence relation in $P\mathcal{M}$:

$$Y_{i+1} := -\mathcal{Q}(\ell\tilde{y} - b) + (E - \mathcal{Q}\ell)Y_i, \qquad (4.1.1\text{-}8)$$

cf. (2.1-7)f. We abbreviate the function $-\mathcal{Q}(\ell\tilde{y} - b)$ by z and the operator $E - \mathcal{Q}\ell$ by k. The recurrence (4.1.1-8) is implemented in the interval screen[†]

$$(IS_N(\mathcal{M}); \diamondsuit, \diamondsuit, \diamondsuit, \diamondsuit)$$

(cf. Section 3.2) in the following way:

$$Z_{i+1} := IS_N z \diamondsuit IS_N(k Z_i), \quad i \geq 0. \qquad (4.1.1\text{-}9)$$

We start with $Z_0 := Y_0 := 0$, and stop when $Z_{k+1} \overset{\circ}{\subset} Z_k$. Calling on Theorem 4 in Section 2.2.1, the existence of a solution y of $\ell y = 0$ is established, and the inclusion $\tilde{y} \diamondsuit Z_{k+1}$ for it has been computed. Moreover, this interval contains only one such solution y. Thus the solution is validated. Note that the validation enables us to conclude that $\mathcal{Q}$ and ℓ are nonsingular, by using Theorem 2 in Chapter 2.

[†] Note that we do not display $\oint$ in this interval-functoid, since we are in the finite case.

In practice, we conduct the iterative computation in $IR_N(\mathcal{M})$ (the isomorphic image of $IS_N(\mathcal{M})$). The recurrence relation employed is

$$W_{i+1} := W_0 + \mathcal{A}(IS_N)*\mathcal{A}(\mathcal{k})*W_i. \qquad (4.1.1\text{-}10)$$

We start with $W_0 := -IR_N(Q(\ell\tilde{y} - b))$ or, if necessary, with the containing interval

$$W_0 := -IR_N(Q(\ell IR_N\tilde{y} - IR_Nb)),$$

and we iterate until $W_{k+1} \overset{o}{\subset} W_k$. In (4.1.1-10) and the equations following it, the arithmetic operations employed are appropriate to the data types displayed, i.e., interval vectors and interval matrices. As before we have $y \in \tilde{y} + Z$ where $Z_k := \Phi*W_k$.

example

We return to the example (4.1.1-3) to illustrate these concepts of validation by means of solution inclusion. Recall that an approximation (cf. (4.1.1-6)) in $S_2(\mathcal{M}) = sp\{1,x,x^2\}$ to the solution of (4.1.1-3) has already been determined. We shall now obtain an inclusion in $IS_2(P\mathcal{M})$ of the solution of (4.1.1-3), and, moreover, by an iterative procedure of the form (4.1.1-9). For purposes of illustration, we begin with the degraded approximation $\tilde{y}(x) = v(3) = (- 1, 1/3)^T$ which is extracted from v in (4.1.1-6). Referring to (2.1-7)f, we take $Q \sim \ell^{-1}(x)$. In particular,

$$Q := \ell^{-1}(3) = \begin{pmatrix} 1 & 6 \\ 1 & 3 \end{pmatrix}^{-1} = \begin{pmatrix} -1 & 2 \\ 1/3 & -1/3 \end{pmatrix}.$$

Then referring to (4.1.1-8)f $(z(x) = - Q(\ell(x)\tilde{y} - b(x))$ and $\mathcal{k} = E - Q\ell)$, we have

$$z = - \begin{pmatrix} -1 & 2 \\ 1/3 & -1/3 \end{pmatrix} \left\{ \begin{pmatrix} 1 & 2x \\ 1 & x \end{pmatrix} \begin{pmatrix} -1 \\ 1/3 \end{pmatrix} - \begin{pmatrix} 1 \\ 0 \end{pmatrix} \right\}$$

$$= \begin{pmatrix} 0 \\ \dfrac{1-x/3}{3} \end{pmatrix}.$$

Then (4.1.1-9) becomes

$$Z_{i+1} = \begin{pmatrix} 0 \\ \dfrac{1-x/3}{3} \end{pmatrix} + \begin{pmatrix} 0 & 0 \\ 0 & 1-x/3 \end{pmatrix} Z_i, \qquad (4.1.1\text{-}11)$$

with $Z_0 := 0$, and where

$$\mathscr{A}(k) = \begin{pmatrix} 0 & 0 \\ 0 & 1-x/3 \end{pmatrix}.$$

Calling $Z_i = (U_i, V_i)^T$, we see that $U_i = 0$, $i = 0,1,2,...$, so that the computation in (4.1.1-1) may be reduced simply to

$$V_{i+1} = \frac{1-x/3}{3} + (1-x/3)V_i, \qquad (4.1.1\text{-}12)$$

with $V_0 = 0$. This iteration is contracting and thus convergent for $0 < x < 6$. This latter interval contains the domain $X = [2,4]$ of our example. The iteration is halted as soon as $V_{i+1} \subset V_i$. Of course, while this is likely to occur even using the rounding in $IS_2(\mathscr{M})$, it is not, in principle, guaranteed to occur. Let $\tilde{V}_{i+1}$ be obtained from V_i exactly as V_{i+1} is, but with all roundings deleted, i.e., to infinite precision. Since all roundings are directed, $\tilde{V}_{i+1} \underset{\circ}{\subset} V_{i+1}$. Thus if the rounded iteration halts because $V_{i+1} \subset V_i$, a fortiori, $\tilde{V}_{i+1} \subset V_i$. Thus appealing to Theorem 4 in Section 2.2.1, *the computation will have demonstrated the existence and uniqueness of a solution* $y = (y_1, y_2)^T$ *along with the bound* $y \in V_{i+1}$ *in this event. That is, the solution will have been validated.*

For our illustrative purposes we may avoid performing this iteration and directly solve (4.1.1-12) for the set V in $IS_2(\mathscr{M})$ in the sense we require:

$$IS_2\left(\frac{1-x/3}{3} + (1-x/3)V \right) \subset V. \qquad (4.1.1\text{-}13)$$

To do this, we use the more appropriate basis $\tilde{\Phi} = \{1, 1-x/3, (1-x/3)^2\}$. Let $s = 1 - x/3$, so that V is sought in the form

$$V := A + Bs + Cs^2.$$

Substitution into (4.1.1-13) now yields

$$IS_2\big((1/3 + A)s + Bs^2 + Cs^3\big) \subset A + Bs + Cs^2. \qquad (4.1.1\text{-}14)$$

We now use *directed Chebyshev rounding*.

Our next step is to compute $IS_2(s^3)$. Noting that $|3s| \leq 1$, we deduce from Table 3.1.2-1 that $S_2(3s^3) = \dfrac{3}{4}(3s)$. Thus,

$$IS_2((3s)^3) = \frac{3}{4}(3s) + \frac{\mathscr{I}}{4}$$

with $\mathscr{I} = [-1, 1]$ and $\sigma = 1$. Finally,

$$IS_2(s^3) = \frac{s}{12} + \frac{\mathscr{I}}{108},$$

so that (4.1.1-14) becomes

$$\frac{\mathscr{I}}{108}C + (\frac{1}{3} + A + \frac{1}{12}C)s + Bs^2 \subset A + Bs + Cs^2.$$

Using the concepts of Section 3.2 and, in particular, using (3.2-4), we see that a sufficient condition for the inclusion displayed here is

$$\frac{\mathscr{I}}{108}C \subset A,$$

$$\frac{1}{3} + A + \frac{1}{12}C \subset B,$$

$$B \subset C.$$

Continuing by setting $B = C$, we come to

$$\mathscr{I}B \subset 108A,$$
$$4 + 12A \subset 11B.$$

From the first of these relations we deduce that $0 \in \mathscr{I} \Rightarrow 0 \in A$, while from the second we deduce that if the diameter of the interval $12A$ is small (compared to 4) then $B \geq 0$.

To continue, we recall that $\rho(X)$ and $\lambda(X)$ denote the right and left end points of an interval X. Now our two inequalities are satisfied if

$$\left.\begin{array}{c} \dfrac{\mathscr{I}}{108}\rho(B) = A \\[2mm] 4 + \dfrac{1}{9}\mathscr{I}\rho(B) = 11B \end{array}\right\} \mathscr{I} = [-1, 1].$$

The last equation here yields $4 + \dfrac{1}{9}\rho(B) = 11\rho(B)$. Then $\rho(B) = 4\dfrac{9}{98} = 0.36735$. Similarly $4 - \dfrac{1}{9}\rho(B) = 11\lambda(B)$ so that $\lambda(B) = 0.35992$. Now the first equation here gives $A = [-1, 1] \times 0.0034014$.

Summarizing, we have

$$V(x) = [-1, 1]0.0034014$$
$$+ [0.35992, 0.36735]\big((1 - x/3) + (1 - x/3)^2\big)$$

and

$$y(x) = \begin{pmatrix} y_1 \\ y_2 \end{pmatrix} \ \epsilon \ \begin{pmatrix} [-1, -1] \\ \dfrac{1}{3} + V(x) \end{pmatrix} =: Z(x).$$

We note in passing that $Z(x)$ is a functional inclusion of the first column of the inverse of the matrix in (4.1.1-3). For purposes of comparison with (the non interval) results shown in the table of (4.1.1-7), we display the following table which is determined from the $V(x)$ just computed. In this table, only two significant digits are retained. In this table $d(V(x))$ denotes the diameter of the interval function $V(x)$ at a point x.

x	$V(x)$	$d(V(x))$	
2	[0.15, 0.17]	0.0071	
3	$[-1, 1] \times 0.0034$	0.0068	(4.1.1-15)
4	$[-0.084, -0.0078]$	0.0071	

Compared to the maximum error bound $|e_2| \leq 0.005$ in (4.1.1-7), we see that the maximum error bound associated with a computed inclusion $d(V) \leq 0.0071$ in (4.1.1-15) is remarkably good. This is especially so when we remember that the former is determined from the exact solution while the latter is determined from the problem data only and through use of an extremely simple basis $\{1,x,x^2\}$.

We continue our study of this example by developing an inclusion through use of a *spline basis* and *directed spline roundings*. For this purpose, we set $X = X_1 \cup X_2$ where $X_1 = [2,3]$ and $X_2 = [3,4]$. We select $\mathcal{M}_i = \mathbb{R}^1[x](X_i)$, $i = 1,2$ for the coefficient space $R_N(\mathcal{M}) = \mathcal{N} \times \mathcal{N}$. The spline space is taken to be $\mathcal{X}\mathcal{K}_1 \times \mathcal{K}_2$ with $P\mathcal{X}\mathcal{K}_1 \times \mathcal{K}_2 = \mathcal{X}P\mathcal{K}_1 \times P\mathcal{K}_2$ (cf. Section 3.1.2). Let us once again choose $\tilde{y} = [-1, 1/3]^T$ as the initial approximation. Then (4.1.1-11) leads to an equation of the form (4.1.1-12); this time in the spline base setting, viz.,

$$V_{i+1} := IS_1\left(\frac{s}{3} + sV_i\right),\qquad (4.1.1\text{-}16)$$

with $V_0 = 0$, and where $s = 1 - x/3$, and

$$V_i \in IS_1(\mathcal{X}^{IR}R_N(P\mathcal{M})) := \mathcal{X}^{II}\mathbb{R}^1[x](X_i) \times {}^{II}\mathbb{R}^1(x)(X_2).$$

We refer to (3.2.2-6)-(3.2.2-12) in proceeding further. Thus with

$$V(x) := V^1(x)\mathcal{X}_1 + V^2(x)\mathcal{X}_2,$$

$$V^1(x) := A + Bs, \quad V^2(x) := C + Ds,$$

$$s = 1 - x/3 \text{ and } s = s\mathcal{X}_1 + s\mathcal{X}_2,$$

(4.1.1-16) is transformed into:

$$V_{i+1}^1\mathcal{X}_1 + V_{i+1}^2\mathcal{X}_2 := IS_{1,1}\left(\frac{s}{3} + sV_i^1\right)\mathcal{X}_1 + IS_{1,2}\left(\frac{s}{3} + sV_i^2\right)\mathcal{X}_2.$$

This leads us to the following pair of recurrences:

$$V_{i+1}^1 := IS_{1,1}\left(\frac{s}{3} + sV_i^1\right), \quad V_0^1 := 0,$$

$$\qquad (4.1.1\text{-}17)$$

$$V_{i+1}^2 := IS_{1,2}\left(\frac{s}{3} + sV_i^2\right), \quad V_0^2 := 0.$$

This iteration is to be continued until the stopping criteria $V_{i+1}^1 \subset V_i^1$ and $V_{i+1}^2 \subset V_i^2$ are met. However, as above and for illustrative purposes, we avoid the literal execution of (4.1.1-17), and solve that equation directly for the fixed points which we seek. Then, using $V^1 = A + Bs$ and $V^2 = C + Ds$ in the stopping criteria for (4.1.1-17), we get

$$IS_{1,1}\left(\frac{s}{3} + As + Bs^2\right) \subset A + Bs, \text{ for } x \in X_1,$$

$$IS_{1,2}\left(\frac{s}{3} + Cs + Ds^2\right) \subset C + Ds, \text{ for } x \in X_2.$$

(4.1.1-18)

$IS_{1,1}$ and $IS_{1,2}$ are taken to be the same namely, directed Chebyshev roundings in X_1 and X_2, respectively. Since $x \in X_1 \iff 3s \in [0,1]$, then

$$IS_{1,1}((3s^2)) = 3s - \frac{1}{8} + \frac{\theta}{9}$$

with $\theta = [-1, 1]$. Equivalently,

$$IS_{1,1}(s^2) = \frac{s}{3} - \frac{\sigma}{12}$$

with $\sigma = [0,1]$.

Since $x \in X_2 \iff s = [-\frac{1}{3}, 0]$, we likewise obtain

$$IS_{1,2}(s^2) = -\frac{s}{3} - \frac{\sigma}{12}.$$

Inserting the values of these roundings into (4.1.1-18), yields

$$-\frac{\sigma}{12}B + \left(\frac{1}{3} + A + \frac{1}{3}B\right)s \subset A + Bs$$

and

$$-\frac{\sigma}{12}D + \left(\frac{1}{3} + C - \frac{1}{3}D\right)s \subset C + Ds.$$

Appealing to (3.2-4), these, in turn, yield

$$-\sigma B = 12A,$$

$$1 + 3A = 2B,$$

and

$$-\sigma D = 12C,$$
$$4 + 12C - 4D = 12D.$$

From the first pair of these equations we compute

$$A := [-\frac{1}{24}, 0] \quad \text{and} \quad B := [-\frac{1}{24}, 0].$$

Thus

$$V^1(x) = [-\frac{1}{24}, 0] + \frac{1}{16}[7, 8](1 - \frac{x}{3}).$$

Similarly from the second pair of equations we compute

$$C := [-0.02638, 0] \quad \text{and} \quad D := [0.201439, 0.3165].$$

Thus

$$V^2(x) := [-0.02638, 0] + [0.201439, 0.3165](1 - \frac{x}{3}).$$

Using $V^1(x)$ and $V^2(x)$, we are able to compute a table analogous to (4.1.1-15):

x	$V(x)$	$d(V(x))$
2	[0.10416, 0.16667]	0.0625
3	[−0.0417, 0]	0.0417
3	[−0.02638, 0]	0.02638
4	[−0.13188, −0.067147]	0.06474

$$\left.\begin{array}{l}\\ \\\end{array}\right\} V^1$$ (4.1.1-19)

$$\left.\begin{array}{l}\\ \\\end{array}\right\} V^2$$

From this table (and the piecewise linearity of V), we can deduce that

$$\| d(V^1(x)) \|_\infty \le 0.0625, \quad \| d(V^2(x)) \|_\infty \le 0.06575.$$

We note that in this case of this particular example, the error associated with the spline rounding (and its four degrees of freedom) is an order of magnitude worse than the error associated with the directed Chebyshev rounding (and its three degrees of freedom). We presume that numbers like 3 and 4 are just too small for appropriate asymptotic relationships to be realized.

4.1.2 The Infinitesimal Case

In the infinitesimal case $\mathscr{L}$ contains at least one operator from among the collection of differential operators, d/dx, integral operators, $\int_0^x$, and linear *shift operators*, $\underset{x}{Sh}^{\alpha x + \beta}$. The lattermost is defined specifically as follows:

$$\underset{y \in \mathscr{M}}{\bigwedge} \underset{x}{Sh}^{\alpha x + \beta} y(x) = y(\alpha x + \beta). \tag{4.1.2-1}$$

Since $\mathscr{L}$ is linear, it is composed of operator products $\ell_1 \circ \ell_2 \ldots \circ \ell_p$ of, say, p operators of the three basic types just enumerated. Recalling our treatment of the operator equations (4.1-1) and (4.1-2) in the finite case, we note that it suffices to produce an isomorphic representation of the three basic infinitesimal operator types, each of which has the form of a matrix that maps $R_N(\mathscr{M}) \rightarrow R_M(\mathscr{M})$. These matrices are denoted by

$$\mathscr{A}\left(\frac{d}{dx}\right), \quad \mathscr{A}\left(\int_0^x\right), \quad \text{and} \quad \mathscr{A}\left(\underset{x}{Sh}^{\alpha x + \beta}\right),$$

respectively. In terms of such matrices, an operator with the form $\mathscr{L} = \ell_1 \ell_2 \ldots \ell_p$ has the representation $\mathscr{A}(\mathscr{L})$ given directly as the following product of matrices:

$$\mathscr{A}(\mathscr{L}) = \mathscr{A}(\ell_1)\mathscr{A}(\ell_2)\ldots\mathscr{A}(\ell_p), \tag{4.1.2-2}$$

where the matrix multiplication here is the conventional one.

We illustrate examples of these matrices in the case of the Taylor basis where $y(x) = \Phi(x) * iy$ with $\Phi(x) = (1, x, \ldots, x^N)$. Recalling the remark following (4.1-1) we only display the leading blocks of interest of the various matrices $\mathscr{A}(\ell)$ to follow.

$$
\mathscr{A}\left(\frac{d}{dx}\right) =
\begin{pmatrix}
0 & 1 & 0 & & . & . & . & . & 0 \\
0 & 0 & 2 & 0 & . & & . & . & 0 \\
& . & & & & . & & & . \\
. & & & & . & & & & . \\
& & & & & & . & & 0 \\
0 & . & . & . & . & . & . & 0 & N \\
0 & . & . & . & . & . & . & 0 & 0
\end{pmatrix}. \quad (4.1.2\text{-}3)
$$

This is an $N' \times N'$ matrix corresponding to a degenerate operator with deflation index $\delta = -1$. (Inflation index is defined in Section 4.1)

$$
\mathscr{A}\left(\int_0^x\right) =
\begin{pmatrix}
0 & & . & . & . & . & 0 \\
1 & 0 & & & & & \\
0 & 1/2 & 0 & & & & . \\
. & & 0 & . & & & . \\
. & & & . & & & . \\
. & & & & . & & \\
& & & & & . & 0 \\
0 & . & . & . & 0 & & 1/(N+1)
\end{pmatrix}. \quad (4.1.2\text{-}4)
$$

This is an $(N' + 1) \times N'$ matrix with inflation index $\delta = 1$.

$$
\mathscr{A}\left(\underset{x}{\overset{ax}{Sh}}\right) =
\begin{pmatrix}
1 & & & & & 0 \\
& \alpha & & & & \\
& & \alpha^2 & & & \\
& & & . & & \\
0 & & & & . & \\
& & & & & \alpha^N
\end{pmatrix}. \quad (4.1.2\text{-}5)
$$

This is an $N' \times N'$ matrix with $\delta = 0$.

The matrices corresponding to

$$
\mathscr{A}\left(\underset{x}{\overset{x+\beta}{Sh}}\right),\ \mathscr{A}\left(\underset{x}{\overset{ax+\beta}{Sh}}\right),\ \mathscr{A}\left(\frac{d^k}{dx^k}\right),\ \text{and}\ \mathscr{A}\left(\left(\int_0^x\right)^k\right)
$$

are given in (4.1.2-6), (4.1.2-7), (4.1.2-8), and (4.1.2-9).

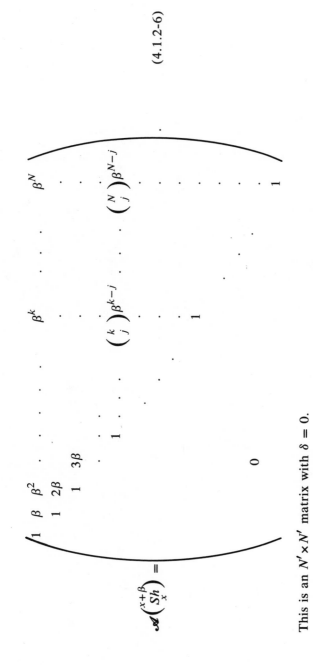

$$\mathscr{A}\!\left(\overset{x+\beta}{S}h_x\right) =
\begin{pmatrix}
1 & \beta & \beta^2 & \cdot & & & & & \beta^k & & & \cdot & & & \beta^N \\
 & 1 & 2\beta & \cdot & & & & & & & & & & & \cdot \\
 & & 1 & 3\beta & \cdot & & & & & & & & & & \\
 & & & \cdot & & & & & & & & & & & \\
 & & & & \cdot & 1 & \cdot & & \binom{k}{j}\beta^{k-j} & \cdot & & \binom{N}{j}\beta^{N-j} \\
 & & & & & & \cdot & & 1 & & & & & \cdot \\
 & & & & & & & \cdot & & & \cdot & \\
 & & & 0 & & & & & & & & & 1 \\
 & & & & & & & & & & & & & & 1
\end{pmatrix}$$

$$(4.1.2\text{-}6)$$

This is an $N' \times N'$ matrix with $\delta = 0$.

$$\mathscr{A}\left(\underset{x}{Sh}^{\,\alpha x+\beta}\right) = \begin{pmatrix} 1 & \beta & \beta^2 & \beta^3 & \cdots & \beta^k & \cdots & \beta^N \\ & \alpha & 2\alpha\beta & 3\alpha\beta^2 & & \cdots & \cdots & \\ & & \alpha^2 & 3\alpha^2\beta & \cdots & & & \\ & & & \alpha^3 & & & & \\ & & & & \ddots & & & \\ & 0 & & & & \binom{k}{j}\alpha^j\beta^{k-j} & \cdots & \binom{N}{j}\alpha^j\beta^{N-j} \\ & & & & & & \ddots & \\ & & & & & & & \alpha^N \end{pmatrix}$$

$$(4.1.2\text{-}7)$$

This is an $N' \times N'$ matrix with inflation index $\delta = 0$.

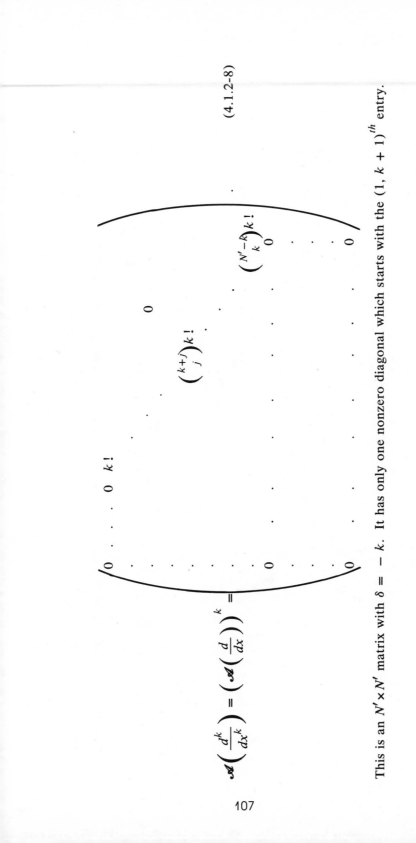

$$\mathscr{A}\left(\frac{d^k}{dx^k}\right) = \left(\mathscr{A}\left(\frac{d}{dx}\right)\right)^k =$$

$$(4.1.2\text{-}8)$$

This is an $N' \times N'$ matrix with $\delta = -k$. It has only one nonzero diagonal which starts with the $(1, k + 1)^{th}$ entry.

$$\mathscr{A}\left(\left(\int_0^x\right)^k\right) = \left(\mathscr{A}\left(\int_0^x\right)\right)^k =$$

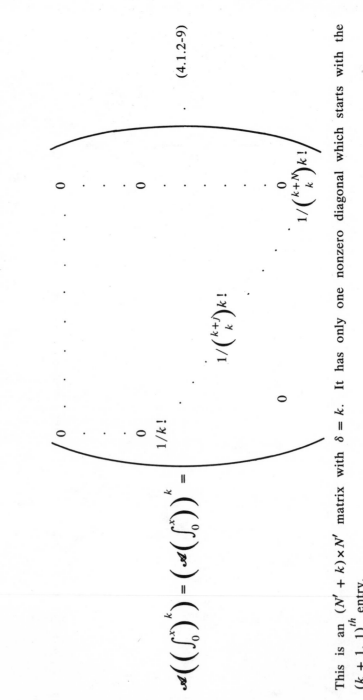

$$\begin{pmatrix} 0 & \cdots & 0 & & & & & 0 & \cdots & 0 \\ & & 1/k! & & & & & & & \\ & & & \ddots & & & & & & \\ & & & & 1/\binom{k+j}{k}k! & & & & & \\ & & 0 & & & \ddots & & & & \\ & & & & & & & 1/\binom{k+N}{k}k! \end{pmatrix}.$$

(4.1.2-9)

This is an $(N' + k) \times N'$ matrix with $\delta = k$. It has only one nonzero diagonal which starts with the $(k + 1, 1)^{th}$ entry.

We also introduce the matrices that represent multiplication by a basis element. The relation

$$x^k v(x) = \Phi * \mathscr{A}(x^k; \bullet) * iv \tag{4.1.2-10}$$

implies that

$$\mathscr{A}(x^k; \bullet) = \begin{pmatrix} 0 & \cdot & \cdot & \cdot & \cdot & \cdot & 0 \\ \cdot & & & & & & \cdot \\ \cdot & & & & & & \cdot \\ \cdot & & & & & & \cdot \\ 0 & & & & & & \cdot \\ 1 & & & & & & \cdot \\ 0 & 1 & & & & & \cdot \\ \cdot & & \cdot & & & & \cdot \\ \cdot & & & \cdot & & & \cdot \\ \cdot & & & & \cdot & & \cdot \\ 0 & & & & & & 1 \end{pmatrix} \qquad \cdot \tag{4.1.2-11}$$

This is an $(N' + k) \times N'$ matrix. It has only one nonzero diagonal which begins with the $(k + 1, 1)^{th}$ entry. Here $\delta = k$.

The more general case of multiplication by an arbitrary polynomial $a(x) = \sum_{k=0}^{N} a_k x^k$ can be constructed by taking linear combinations of the matrices $\mathscr{A}(x^k; \bullet)$. In particular, writing

$$a(x)v(x) = \Phi * \mathscr{A}(a; \bullet) iv,$$

we have

$$\mathscr{A}(a; \bullet) = \sum_{k=0}^{p} a_i \mathscr{A}(x^i; \bullet).$$

$\mathscr{A}(a; \bullet)$ is an $(N' + p + 1) \times N'$ matrix. It is a Toeplitz matrix consisting of the N' nonzero diagonals as indicated in (4.1.2-12). Its deflation index $\delta = N$. Explicitly,

$$\mathcal{A}(a;\bullet) = \begin{pmatrix} a_0 & & & & \\ a_1 & a_0 & & 0 & \\ a_2 & & \cdot & & \\ \cdot & & & \cdot & \\ \cdot & & & & \cdot \\ \cdot & & & & a_0 \\ a_N & & & & a_1 \\ & a_N & & & a_2 \\ & & & & \cdot \\ & & 0 & \cdot & \cdot \\ & & & & a_N \end{pmatrix} \qquad (4.1.2\text{-}12)$$

example

We consider the following integral equation as a sample problem.

$$y \equiv \ell y := \sigma x^3 y + \int_0^x t^2 y(\alpha t + \beta)dt + x^4.$$

Here ℓ is represented by the following sequence:

$$u_1 = \sigma x^3 y \equiv: \ell_1 y,$$

$$u_2 = y(\alpha x + \beta) \equiv: \ell_2 y,$$

$$u_3 = x^2 u_2 \equiv: \ell_3 u_2,$$

$$u_4 = \int_0^x u_3 dt \equiv: \ell_4 u_3,$$

$$u_5 = u_1 + u_4 \equiv: \ell_5 u_4 = \ell_1 y + \ell_4 \ell_3 \ell_2 y.$$

Then

$$y = \ell y := (\ell_1 + \ell_4 \ell_3 \ell_2) y + x^4,$$

so that

$$\mathscr{A}(\ell)\bullet = \big(\mathscr{A}(\ell_1) + \mathscr{A}(\ell_4)\mathscr{A}(\ell_3)\mathscr{A}(\ell_2)\big)\bullet + ix^4. \qquad (4.1.2\text{-}13)$$

In fact we have already computed these $\mathscr{A}(\ell_j)$, $j = 1,2,3,4$, so that combining, we may write $\mathscr{A}(\ell)*iv$ as

$$
\left(
\begin{array}{cccccccc}
0 & \cdot & \cdot & \cdot & \cdot\;\cdot & \cdot & 0 \\
0 & \cdot & \cdot & \cdot & \cdot\;\cdot & \cdot & 0 \\
0 & \cdot & \cdot & \cdot & \cdot\;\cdot & \cdot & 0 \\
\sigma+1/3 & \beta/3 & \beta^2/3 & \beta^3/3 & \cdot\;\;\cdot & \cdot & \beta^N/3 \\
0 & \sigma+\alpha/4 & 2\alpha\beta/4 & 3\alpha\beta^2/4 & \cdot\;\;\cdot & \cdot & N\alpha\beta^{N-1}/4 \\
 & & & \cdot & \cdot & \cdot\;\cdot & \cdot \\
 & & & & \cdot & \cdot\;\cdot & \cdot \\
 & & 0 & & & \cdot\;\cdot & \cdot \\
 & & & & & \cdot\;\cdot & N\alpha^{N-1}\beta/N \\
 & & & & & & \sigma+\alpha^N/N+1 \\
 & & & & & & 0
\end{array}
\right)
\bullet iv +
\left(
\begin{array}{c}
0 \\ 0 \\ 0 \\ 0 \\ 1 \\ 0 \\ \cdot \\ \cdot \\ \cdot \\ \cdot \\ 0
\end{array}
\right).
$$

Here $\delta = 3$.

This concludes presentation of the example.

Now let us turn to the question of computing a solution of an equation, viz., $y = \ell y$ (resp. $\mathscr{L}y = 0$). We proceed by rounding onto the screen $S_N(\mathscr{M})$ and solving a modified problem along the lines dealt with at the beginning of Chapter 4, viz.,

$$v = S_N \ell v \quad (\text{resp. } S_N \mathscr{L}v = 0),$$

or

$$iv = i(S_N \ell)iv \quad (\text{resp. } i(S_N \mathscr{L})iv = 0),$$

or

$$iv = \mathscr{A}(S_N)\mathscr{A}(\ell)*iv \quad (\text{resp. } \mathscr{A}(S_N)\mathscr{A}(\mathscr{L})*iv = 0). \qquad (4.1.2\text{-}14)$$

The last equation represents a linear system in explicit resp. implicit form. iv is the unknown vector being sought. $i(S_N\ell) = \mathscr{A}(S_N)\mathscr{A}(\ell)$ is an affine operator, an example of which is given in the sample problem just treated. Likewise $i(S_N \mathcal{L}) = \mathcal{A}(S_N)\mathcal{A}(\mathcal{L})$ is an affine operator as well.

Employing the solution iv of (4.1.2-2), we generate

$$\tilde{y}(x) = \Phi(x)*iv$$

as an approximation of the solution y. iv itself can be determined from (4.1.2-14) by any standard numerical method for solving linear systems.

example

As an example of the solution procedure under discussion, we consider the model functional equation

$$\mathcal{L}y := 1 + \int_0^x y - y \equiv \ell y - r = 0,$$

where

$$\ell y := \int_0^x y - y, \; r := -1,$$

and on the domain $X := [-1, 1]$.

We shall set $N = 2$ and employ the basis $\Phi_N = \{1, x, x^2\}$. Then $M = 3$. With the notation of (4.1-1) our equation can now be written in the form

$$\mathcal{A}(S_2)(\mathcal{A}(\ell)iv - ir) = 0.$$

Upon identifying iv with v and ir with r, this leads to

$$\mathcal{A}(S_2)\mathcal{A}(\ell)v = \mathcal{A}(S_2)r. \tag{4.1.2-15}$$

Now using *Chebyshev rounding*, (3.1.2-1)-(3.1.2-3), and Table 3.1.2-1, we have

$$\hat{\mathcal{A}}(S_2) = \begin{vmatrix} 0 & -1/8 & . & . & . \\ 3/4 & 0 & . & . & . \\ 0 & 1 & . & . & . \end{vmatrix}.$$

Similarly, using (4.1.2-2) and (4.1.2-4), we find

$$\mathscr{A}(\ell) = \begin{pmatrix} -1 & 0 & 0 & \cdot & \cdot & \cdot \\ 1 & -1 & 0 & \cdot & \cdot & \cdot \\ 0 & 1/2 & -1 & \cdot & \cdot & \cdot \\ 0 & 0 & 1/3 & \cdot & \cdot & \cdot \\ \cdot & & & \cdot & & \\ \cdot & & & & \cdot & \\ \cdot & & & & & \cdot \end{pmatrix}.$$

Now using (4.1-3) and (4.1-4) and noting that $\overset{\wedge}{\mathscr{A}}(\ell)$ is just the block consisting of the first three rows of $\mathscr{A}(\ell)$ here, we have

$$\Gamma := \mathscr{A}(S_2)\mathscr{A}(\ell) = \begin{vmatrix} -1 & 0 & 0 \\ 1 & -1 & 0 \\ 0 & 1/2 & -1 \end{vmatrix} + \begin{vmatrix} 0 & 0 & 0 \\ 0 & 0 & 1/4 \\ 0 & 0 & 0 \end{vmatrix}$$

$$= \begin{vmatrix} -1 & 0 & 0 \\ 1 & -1 & 1/4 \\ 0 & 1/2 & -1 \end{vmatrix}.$$

Similarly

$$-\bar{r} := -\mathscr{A}(S_2)ir = \begin{vmatrix} -1 \\ 0 \\ 0 \end{vmatrix}.$$

Then (4.1.2-15) becomes

$$\Gamma v = \bar{r}, \qquad\qquad (4.1.2\text{-}16)$$

so that $v = (1, 8/7, 4/7)^T$.

Collecting, we have

$$v(x) = \Phi_2 * v = 1 + \frac{8}{7}x + \frac{4}{7}x^2.$$

Using this, we produce Table 4.1.2-1.

x	y	v	$\|y-v\|$	$\|\dfrac{y-v}{y}\|$
1	e	$19/7 \sim 2.7142857$	0.004	0.00147
-1	e^{-1}	$3/7 \sim 0.42857143$	0.06	0.165
$\dfrac{1}{3}$	$e^{1/3}$	1.444444	0.049	0.050

Table 4.1.2-1

Let us continue with this example, but let us employ a *spline basis* and a *spline rounding.* We partition X into $X_1 = [-1, -1/3]$, $X_2 = [-1/3, 1/3]$, and $X_3 = [1/3, 1]$, and we take $\mathscr{X}^{\mathscr{N}_1 \times \mathscr{N}_2 \times \mathscr{N}_3}$ with $\mathscr{N}_i = \mathbb{R}^1[x](X_i)$, $i = 1,2,3$ as the approximating subspace. That is, we consider approximations $v(x)$ of the form

$$v(x) = v_1 \mathscr{X}_1 + v_2 \mathscr{X}_2 + v_3 \mathscr{X}_3$$

with $v_i \in \mathscr{N}_i$, $i = 1,2,3$. In particular, setting $\mathscr{X} = (\mathscr{X}_1, \mathscr{X}_2, \mathscr{X}_3)$ and

$$t = \begin{pmatrix} v_1 \\ v_2 \\ v_3 \end{pmatrix} := \begin{pmatrix} a_1 + b_1(x + \frac{1}{3}) \\ a_2 + b_2 x \\ a_3 + b_3(x - \frac{1}{3}) \end{pmatrix} := \begin{pmatrix} \tilde{a}_1 + b_1 x \\ \tilde{a}_2 + b_2 x \\ \tilde{a}_3 + b_3 x \end{pmatrix},$$

we take

$$v = \mathscr{X} * t.$$

Now

$$\int_0^x v = \mathscr{X} * \mathscr{A}\left(\int_0^x\right) * t.$$

Referring to (3.1.2-8)-(3.1.2-10), we compute $\mathscr{A}\left(\int_0^x\right)$ as follows:

$$\int_0^x v = \int_0^x \mathscr{X} * t$$

$$= \int_0^x v_1 \mathscr{X}_1 + \int_0^x v_2 \mathscr{X}_2 + \int_0^x v_3 \mathscr{X}_3$$

$$= \begin{cases} \displaystyle\int_{-1/3}^x v_1 + \int_0^{-1/3} v_2, & x \in (-1,-\tfrac{1}{3}) \\[12pt] \displaystyle\int_0^x v_2, & x \in (-\tfrac{1}{3},\tfrac{1}{3}) \\[12pt] \displaystyle\int_0^{1/3} v_2 + \int_{1/3}^x v_3, & x \in (\tfrac{1}{3},1) \end{cases}$$

$$= \mathscr{X}_1 \left(\int_{-1/3}^x v_1 + \int_0^{-1/3} v_2 \right) + \mathscr{X}_2 \int_0^x v_2 + \mathscr{X}_3 \left(\int_0^{1/3} v_2 + \int_{1/3}^x v_3 \right)$$

$$:= (\mathscr{X}_1, \mathscr{X}_2, \mathscr{X}_3) \mathscr{A}\left(\int_0^x \right) \begin{pmatrix} v_1 \\ v_2 \\ v_3 \end{pmatrix}.$$

Thus $\mathscr{A}(\int)$ is a matrix of operators each of which acts on $\mathscr{X}_i$:

$$\mathscr{A}\left(\int_0^x \right) = \begin{pmatrix} \displaystyle\int_{-1/3}^x & \displaystyle\int_0^{-1/3} & 0 \\[12pt] 0 & \displaystyle\int_0^x & 0 \\[12pt] 0 & \displaystyle\int_0^{1/3} & \displaystyle\int_{1/3}^x \end{pmatrix}$$

Hence

$$\mathscr{A}(\ell) = \mathscr{A}\left(\int_0^x \right) - E,$$

where E is the 3×3 identity matrix.

Referring to (3.1.2-15), we see that $\mathscr{A}(S_1)$ is the following diagonal matrix of rounding operators:

$$\mathcal{A}(S_1) = \begin{pmatrix} S_{1,1} & 0 & 0 \\ 0 & S_{1,2} & 0 \\ 0 & 0 & S_{1,3} \end{pmatrix},$$

where since $N = 1$ and $M = 2$, we have

$$S_{1,i}: \mathbb{R}^2[x](X_i) \rightarrow \mathbb{R}^1[x](X_i), \ i = 1,2,3.$$

Thus we are to apply all operators blockwise on each of the components of the spline basis.

Set $u = (a_1, b_1, a_2, b_2, a_3, b_3)^T$. Then employing (3.1.2-1), (3.1.2-3), (3.1.4), (4.1.2-2), (4.1.2-4), and (4.1-4), v is determined through the following equation for its coefficients:

$$\Gamma u = (-1, 0, -1, 0, -1, 0)^T, \qquad (4.1.2-17)$$

where

$$\Gamma = \begin{pmatrix} -1 & -1/36 & | & -1/3 & 1/18 & | & 0 & 0 \\ 1 & -4/3 & | & 0 & 0 & | & 0 & 0 \\ - & - & - & | & - & - & - & - & |- & - & - & - \\ 0 & 0 & | & -1 & 1/36 & | & 0 & 0 \\ 0 & 0 & | & 1 & -1 & | & 0 & 0 \\ - & - & - & | & - & - & - & - & |- & - & - & - \\ 0 & 0 & | & 1/3 & \cdot 1/18 & | & -1 & -1/36 \\ 0 & 0 & | & 0 & 0 & | & 1 & 2/3 \end{pmatrix}.$$

Then

$$u = \left(\frac{3640}{5292}, \frac{910}{1764}, \frac{36}{35}, \frac{3528}{2625}, \frac{1764}{875} \right),$$

and so,

$$v_1 = \frac{3640}{5292} + \frac{910}{1764}(x + \frac{1}{3}), \quad -1 \le x \le -\frac{1}{3},$$

$$v_2 = \frac{36}{35}(x + 1), \qquad\qquad |x| \le \frac{1}{3},$$

$$v_3 = \frac{3528}{2625} + \frac{1764}{875}(x + \frac{1}{3}), \qquad \frac{1}{3} \le x \le 1.$$

From this in turn we produce the following table.

x	y	v	$|y - v|$
1	e	2.688	.03
-1	e^{-1}	0.3439	.02
$\frac{1}{3}$	$e^{1/3}$	1.3714	.02

Comparing this table with Table 4.1.2-1, we see that the quality of the approximations produced by solving (4.1.2-16) or (4.1.2-17) are nearly the same. While the corresponding systems are of third and sixth orders, respectively, the coefficient matrix of the former is full while that of the latter is sparse.

VALIDATION (automatic error control)

Let $\tilde{y} \in S_N(\mathcal{M})$ be an approximation of the solution y of the equation $y = \ell y$ being sought. For example, $\tilde{y}$ could be obtained by the process leading to (4.1.1-8). We next calculate a correction Y, an interval, so that $y \in \tilde{y} + Y$. To do this, we use the following recurrence relation

$$Y_{i+1} := \ell\tilde{y} - \tilde{y} + \ell Y_i, \quad i \ge 0,$$

with $Y_0 = 0$.

We implement this recurrence in the interval screen (more precisely, interval functoid) $(IS_N(\mathcal{M}); IS_N(\Omega))$. Abbreviating $\ell\tilde{y} - \tilde{y}$ by z and setting $Z_0 = 0$, this implementation is given by (analogous to (4.1.1-9))

$$Z_{i+1} = IS_N z \,\diamondplus\, IS_N \ell Z_i, \quad i \ge 0. \qquad (4.1.2\text{-}18)$$

We iterate this recurrence until $Z_{k+1} \overset{\circ}{\subset} Z_k$. Calling on Theorem 4 in Section 2.2.1, we have that $y \in \widetilde{y} \oplus Z_{k+1}(x)$. The solution y will then have been validated. That is, an inclusion of the solution y being sought will have been produced along with a demonstration of its existence and uniqueness in the interval $\widetilde{y} \oplus Z_{k+1}$.

In practice, we conduct the iterative computation in $IR_N(\mathcal{M})$ (the isomorphic image of $IS_N(\mathcal{M})$). The recurrence relation employed is (analogous to (4.1.1-10))

$$W_{i+1} := W_0 + \mathcal{A}(IS_N) * \mathcal{A}(\mathcal{L}) * W_i, \quad i \geq 0. \tag{4.1.2-19}$$

We start with $W_0 := \mathcal{A}(IS_N)*(\mathcal{A}(\mathcal{L}) - E)*IR_N(\widetilde{y})$, and iterate until $W_{k+1} \overset{\circ}{\subset} W_k$. When this occurs $Z_{k+1} := \Phi*W_{k+1}$ and $y \in \widetilde{y} + Z_{k+1}$. For implementation of $\mathcal{A}(\mathcal{L})$, we may use its decomposition in the product form of (4.1.2-2). For the individual components of this product, we may use the matrices (4.1.2-3)-(4.1.2-11). Note that in the present context of validation the matrix (4.1.2-12) representing multiplication by a polynomial may not be used. This prohibition is made since direct use of the product by polynomial matrix does not generally supply the isotoney property required by the operations in an interval functoid.

example
As an example of this process we return to our model functional equation

$$\mathcal{L}y := 1 + \int_0^x y\,dx - y = 0.$$

Referring to (2.1-7), we seek a form of this equation which is in some sense practical for iteration. An obvious candidate is

$$y = 1 + \int_0^x y\,dx = \ell y. \tag{4.1.2-20}$$

We round this problem into the subspace $I\!I\!R^2[x][-1, 1]$, and direct our attention to determining a fixed point in this subspace of the following recurrence:

$$Y_{i+1} = IS_2(1 + \int_0^x Y_i(x)dx), \quad i \geq 0,$$

with $Y_0 = 0$.

The fixed point is characterized by the stopping criterion $Y_{k+1} \overset{\circ}{\subset} Y_k$. From Theorem 4 in Section 2.2.1 the existence of such a Y_{k+1} implies that a solution $y(x) \in Y_{k+1}$ of (4.1.2-20) exists and is unique. That is, the solution is validated.

The actual iteration performed on a computer is conducted in terms of representations of the Y_i of the form $V(x) = A + Bx + Cx^2$ where $A, B, C \in \mathbb{R}$. In particular, we have the isomorphic recurrence

$$iV_{i+1}(x) = \begin{vmatrix} A \\ B \\ C \end{vmatrix}_{i+1} := I\Gamma \begin{vmatrix} A \\ B \\ C \end{vmatrix}_i + \begin{vmatrix} 1 \\ 0 \\ 0 \end{vmatrix}$$

$$= I\Gamma iV_i(x) + \begin{matrix} 1 \\ 0 \\ 0 \end{matrix}$$

in $R_2(\mathbb{R}^2[x][-1, 1])$. Here

$$I\Gamma := \mathscr{A}(IS_2)\mathscr{A}\left(\int_0^x\right).$$

As in Section 4.1.1, we avoid the actual process of iteration, and proceed to solve the following equation, which expresses the fixed point property directly:

$$A + Bx + Cx^2 \supset IS_2(I + Ax + \frac{1}{2}Bx^2 + \frac{1}{3}Cx^3). \qquad (4.1.2\text{-}21)$$

Since

$$IS_2(x^3) = \frac{3}{4}x + \frac{[-1, 1]}{4},$$

(4.1.2-21) becomes

$$A + Bx + Cx^2 \supset 1 + \frac{\theta}{12}C + (A + \frac{1}{4}C)x + \frac{1}{2}Bx^2,$$

with $\theta = [-1, 1]$. From this in turn we derive

$$A \supset 1 + \frac{\theta}{12}C,$$

$$B \supset A + \frac{1}{4}C,$$

$$C \supset \frac{1}{2}B.$$

We note incidentally that $I\Gamma$ may be read off from these equations, viz.,

$$I\Gamma = \begin{pmatrix} 0 & 0 & \frac{\theta}{12} \\ 1 & 0 & \frac{1}{4} \\ 0 & \frac{1}{2} & 0 \end{pmatrix}.$$

To solve the inclusion relations, it suffices to apply an elimination process in which inclusion is regarded as equality, the most stringent inclusion. Eliminating C from the inclusion equations above and then eliminating A, we obtain

$$B \supset \frac{\theta}{24}B + \frac{1}{8}B + 1.$$

From this in turn (and assuming that $B > 0$), we get

$$\lambda(B) \leq -\frac{\rho(B)}{24} + \frac{\lambda(B)}{8} + 1,$$

$$\rho(B) \geq \frac{\rho(B)}{24} + \frac{\rho(B)}{8} + 1.$$

Thus

$$B = \left[\frac{38}{35}, \frac{6}{5}\right],$$

and correspondingly,

$$A = \frac{[19,21]}{20}, \quad C = \left[\frac{19}{35}, \frac{3}{5} \right].$$

Thus we have demonstrated that

$$y(x) = e^x \in V(x) = 1 + B\left(\frac{\theta}{24} + x + \frac{1}{2}x^2 \right)$$

$$= 1 + \left[\frac{38}{35}, \frac{6}{5} \right]\left(\frac{[-1,\ 1]}{24} + x + \frac{1}{2}x^2 \right).$$

Correspondingly, we have the following associated inclusions:

$$e \in V(1) = [2.583,\ 2.85]$$

and

$$e^{-1} \in V(-1) = [0.1857,\ 0.45].$$

The low quality of the error bound is only a reflection of the primitive nature of the basis employed in the example.

We stress once again that the point to note is that the method of computation itself using nothing more than the functional equation as data produces bounds for the solution that moreover estimate both of the errors of computation at once; the discretization error (i.e., error of rounding $IS_2\colon P\mathcal{M} \to I\!\!R^2[x]([-1, 1]))$ and the rounding error of the actual computing performed. Moreover, it may be shown by methods of approximation theory that the quality of bounds of the type we are computing may be increasingly improved by increasing the index N.

We conclude this section with two examples of the computation of an inclusion. The first example shows how the quality of the inclusion may be made arbitrarily good for the exponential function. This is a critical feature for evaluating arithmetic expressions to full accuracy (cf. [5], [14]). The second example demonstrates the employment of splines for producing an inclusion.

example 1:

Consider the model initial-value problem

$$y' = y, \quad y(0) = 1. \tag{4.1.2-22}$$

Set $y(x) = w^t(x)$ with t a parameter. Then (4.1.2-22) becomes

$$w'(x) = \frac{1}{t} w(x), \quad w(0) = 1. \tag{4.1.2-23}$$

To approximate w, we take the ansatz $w = a + bx$. Then

$$y(x) \sim \left(1 + \frac{2x}{2t-1} \right)^t \quad \text{for all } t > 0.$$

As a way of particularizing t, let us employ the methods of this section (see the example following (4.1.2-14)). That is, we compute an inclusion of y and seek to minimize the diameter of the inclusion with respect to t. This procedure will illustrate the production of a guarantee for the value of solution as the guarantee is refined with dynamically increasing accuracy.

Proceeding as in the cited example (converting (4.1.2-23) to an integral equation, etc.), we find that

$$a = 1, \quad b = \left[\frac{1}{t}, \frac{1}{t-1} \right].$$

Hence

$$y(x) \in V_t(x) = (1 + \left[\frac{1}{t}, \frac{1}{t-1} \right] x)^t, \quad t > 0, \quad 0 \le x \le 1.$$

Then we have the following bound for $d(V_t(x))$, the diameter of the interval $V_t(x)$ (cf. [1]):

$$d(V_t(x)) \le t \left(1 + \frac{x}{t-1} \right)^{t-1} \frac{x}{t(t-1)}$$

$$= \left(1 + \frac{x}{t-1} \right)^{t-1} \frac{x}{t-1}.$$

This bound vanishes as $t \to \infty$. However, since (when t is an integer and) from the arithmetic point of view, the number of multiplications required to calculate this bound increases indefinitely with t, then in practice, t is necessarily bounded.

For $t = 4$, for example, we have

$$y(x) = e^x \in \left(1 + [\tfrac{1}{4}, \tfrac{1}{3}]x \right)^4,$$

so that

$$e \in [2.44, 3.16].$$

For $t = 8$, we have correspondingly that

$$e \in [2.56, 2.911].$$

The apparently poor progress in improving these bounds with increasing t is due to the linear ansatz for w. Taking the ansatz $w = 1 + ax + bx^2$ instead, we find

$$y(x) = e^x \sim \left(1 + ((2t - 1)x + x^2)\frac{8}{8t(2t - 1) + 1} \right)^t,$$

and, in particular,

$$e \sim \left(1 + \frac{16t}{8t(2t - 1) + 1} \right)^t.$$

The corresponding system characterizing the inclusion reads

$$ta \subset 1 - \frac{\sigma}{4}b,$$

$$2tb \subset a + b.$$

This leads to

$$a = \left[1 - \frac{1}{4t(2t-1)}, 1 \right] \frac{1}{t^2(2t-1)},$$

$$b = \frac{a}{t}.$$

The following table displays results of these computations.

t	$e \in (1 + 2tb)^t$	$d(V_t(1))$
2	[2.6859, 2.7778]	0.092
4	[2.71098, 2.7327]	0.023
8	[2.71649, 2.72184]	0.0054

We expect $\| d(V_t(x)) \| = O(1/t^N)$, where N is the order of the polynomial in the ansatz for w.

Using (4.1.2-23) and these methods for approximating the exponential on $0 \leq x \leq 1$ supplies not only an inclusion but an inclusion whose accuracy may be made arbitrarily good by increasing t. The method is one of dynamic accuracy with the work involved being proportional to the accuracy required (i.e., to t). For accomplishing so many objectives relative to the evaluation of the exponential, the work involved is seemingly modest.

example 2:

This example shows how splines are used in the production of an inclusion. We employ the model problem (4.1.2-20). We partition $X = [-1, 1]$ into $X_1 = [-1, 0]$ and $X_2 = [0, 1]$, and we take the simple coefficient space $IR_N(P\mathcal{M}) := \mathcal{N} \times \mathcal{N}$ with $\mathcal{N} = II\,\mathbb{R}^2[x]$. Hence we work in the directed screen

$$\mathcal{X}^{II\,\mathbb{R}^2[x] \times II\,\mathbb{R}^2[x])}.$$

Proceeding directly, we take the ansatz

$$V_1 = A_1 + B_1 x + C_1 x^2, \quad x \in X_1,$$

$$V_2 = A_2 + B_2 x + C_2 x^2, \quad x \in X_2.$$

Corresponding to (4.1.2-20), we have the following conditions for fixed points in the spline basis:

$$A_i + B_i x + C_i x^2$$

$$(4.1.2\text{-}24)$$

$$\supset IS_{2,i}(1 + A_i x + \frac{1}{2} B_i x^2 + \frac{1}{3} C_i x^3), \quad i = 1,2.$$

For the roundings following (3.2.2-12), we have similarly to (4.1.1-16)f

$$IS_{2,1}(x^3) = \frac{48}{32} x^2 - \frac{18}{32} x + \frac{[0,1]}{16},$$

$$IS_{2,2}(x^3) = -\frac{48}{32} x^2 - \frac{18}{32} x - \frac{[0,1]}{16}.$$

Combining, we derive

$$A_1 \supset 1 + \frac{\sigma_1}{48} C_1, \qquad A_2 \supset 1 - \frac{\sigma_2}{48} C_2,$$

$$B_1 \supset A_1 - \frac{3}{16} C_1, \qquad B_2 \supset A_2 - \frac{3}{16} C_2, \quad (4.1.2\text{-}25)$$

$$C_1 \supset \frac{1}{2} B_1 + \frac{1}{2} C_1, \qquad C_2 \supset \frac{1}{2} B_2 - \frac{1}{2} C_2,$$

for all $\sigma_i = X_i$, $i = 1,2$.

As in the previous example, the isomorphic representation $I\Gamma$ of the rounded operator $\ell = \int_0^x$ may be read off from the relations here.

$$I\Gamma = \mathscr{A}\left(\mathscr{A}(IS_2) \mathscr{A}\left(\mathscr{A}(\int_0^x) \right) \right),$$

where

$$
I\Gamma = \left(\begin{array}{ccc|ccc}
0 & 0 & \dfrac{\sigma}{48} & & & \\
1 & 0 & -\dfrac{3}{16} & & 0 & \\
0 & \dfrac{1}{2} & \dfrac{1}{2} & & & \\
\hline
& & & 0 & 0 & -\dfrac{\sigma}{48} \\
& 0 & & 1 & 0 & -\dfrac{3}{16} \\
& & & 0 & \dfrac{1}{2} & -\dfrac{1}{2}
\end{array}\right).
$$

Setting $T = (A_1, B_1, C_1, A_2, B_2, C_2)^T$, the relations for the fixed point are collectively written as

$$
T \supset I\Gamma T + (1,0,0,1,0,0)^T.
$$

To solve the first set of relations in (4.1.2-25), we note that the last relation there implies that $B_1 = C_1$. (Refer to the comment following (4.1.2-21) concerning replacing inclusion by equality for the solution process.) Then, replacing C_1 by B_1 in the first two relations and then combining them to eliminate A_1, we obtain

$$
B_1 \supset \frac{\sigma}{48} B_1 - \frac{3}{16} B_1 + 1, \quad \text{for all } \sigma_1 \in [-1, 0].
$$

Then,

$$
\lambda(B_1) \leq -\frac{1}{48}\rho(B_1) - \frac{3}{16}\rho(B_1) + 1 = -\frac{5}{24}\rho(B_1) + 1
$$

$$
\rho(B_1) \geq \frac{3}{16}\lambda(B_1) + 1.
$$

From these two equations, in turn, we find the interval $B_1 (= C_1)$. In particular,

$$
B_1 = C_1 = [0.83862068, 0.86068967].
$$

Combining this with (4.1.2-23), we find

$$A_1 = [1.0, \ 1.0179311].$$

To solve for A_2, B_2, and C_2 we eliminate A_2 resp. B_2 from the second resp. third equation to arrive at an inclusion for C_2

$$C_2 \supset -\frac{[57, \ 58]}{96} C_2 + \frac{1}{2}.$$

Hence,

$$C_2 = [0.30862943, \ 0.31675128 \ ],$$
$$A_2 = [0.99340101, \ 1.0 \ ],$$
$$B_2 = [0.93401014, \ 0.94213199 \ ].$$

Summarizing, we have

$$V_1(x) = 1 + B_1\left(\frac{[0, \ 1]}{48} + x + x^2\right),$$

$$V_2(x) = A_2 + B_2 + C_2 x^2.$$

Correspondingly, we have the associated inclusions

$$e \in V_1(1) = [2.6772412, \ 2.7393105 \ ],$$

$$e^{-1} \in V_2(-1) = [0.35989845, \ 0.38274118 \ ],$$

$$\|d(V_1(x))\| \ \leq \ d(A_1) + d(B_1) + d(C_1) = 0.0621,$$

$$\|d(V_2(x))\| \ \leq \ d(A_2) + d(B_2) + d(C_2) = 0.0285.$$

Using the approximations $\|V_1\| = e$, $\|V_2\| = 1$, we find the relative errors

$$\frac{\|d(V_1)\|}{\|V_1\|} \leq 0.0285, \quad \frac{\|d(V_2)\|}{\|V_2\|} \leq 0.0285,$$

which displays a symmetry and suggests a constant relative error.

4.2 METHODS FOR NONLINEAR FUNCTIONAL EQUATIONS

The concepts and methodology introduced in Section 4.1 are valid and are generally applicable to the nonlinear case. This is especially so when local linearizations are used, such as Newton's method. A salient difficulty of the nonlinear case is the determination of which one of a set of solutions is to be sought. The corresponding features, such as starting element selection and local convergence, which arise for dealing with this difficulty, will have their counterparts in our considerations.

We shall first treat the finite case in Section 4.2.1 and then the infinitesimal case in Section 4.2.2. Our treatment will consist of a sequence of examples that demonstrate the relevant ideas.

4.2.1 The Finite Case

Two typical functional equations which exemplify the finite case are

$$y(2x) = 2y^2(x) - 1, \quad 0 \le x \le 1, \qquad (4.2.1\text{-}1)$$

with $y(1) = 0$, and

$$u(x + y) = u(x)u(y), \quad u(0) = 1, \quad u(1) = e. \qquad (4.2.1\text{-}2)$$

The cosine and the exponential are solutions of these two problems, respectively. Thus such problems are of particular interest in the context of computers equipped with standard functions or with packages for functions, since the cosine and the exponential are canonical. This is especially so if such functions are supplied with arbitrarily specifiable accuracy.

The need for such accuracy is exposed by ill-conditioned problems in which such basic functions occur. For example, consider the following problem: determine a zero $x \sim 1.4$ of the equation $f(x) = 0$, where

$$\begin{aligned}
f(x) = \ &\cos\,(67872320568x^3 - 95985956257x^2 \\
&- 135744641136x + 191971912515) \qquad (4.2.1\text{-}3) \\
&- 0.5401873.
\end{aligned}$$

The computation of this zero requires residue iteration, employing the exact scalar product. This in turn requires evaluation of the cosine with an accuracy that increases dynamically with the steps of the residue iteration. A practical way to deliver the cosine is through a defining system for which repeated iteration supplies the accuracy required. Such a defining system is (4.2.1-1) augmented by some additional constraints. The functional equation in (4.2.1-1) is algebraically nonlinear and is thus exactly definable in a computer. Thus the solution of this functional equation may be approximated to any desired accuracy by the residue iteration method.

In particular the cosine needed for the problem (4.2.1-3) may be computed using the methodology of this so-called finite case. This is done in the example that follows, illustrating this methodology. In fact it is preferable to compute $f(x)$ in (4.2.1-3) directly as a solution of a differential equation. This treatment corresponds to a so-called infinitesimal case. This latter case is treated in Section 4.2.2 (see (4.2.2-15)).

example

Let us apply our methodology to the functional equation (4.2.1-1) which we take into the form

$$y(x) = 2y^2\left(\frac{x}{2}\right) - 1, \quad y(1) = 0. \qquad (4.2.1\text{-}4)$$

The solution of (4.2.1-4) is $\cos\frac{\pi}{2}x$. To simplify computation we choose $X = [0, 1]$, exploiting the symmetry of the solution, $y(-x) = y(x)$.

Our approximations $v(x)$ to $y(x)$ are found in our customary manner: seek $v \in S_N(\mathcal{M})$, etc.

Employing the Chebyshev rounding, we find the following results:

$$N = 2: \quad v(x) = 0.9817286(1 - x^2),$$
$$N = 4: \quad v(x) = 0.9997880539 - 1.233233x^2 + 0.2334417x^4.$$

Values and errors associated with these polynomials are computed at

$x = 1/2$ and are shown in the table

N	$v(1/2)$	$\left\| \dfrac{y(1/2)-v(1/2)}{y(1/2)} \right\|$
2	0.7363	0.0413
4	0.7060700	0.00147

By applying the functional equation to itself recursively, the value of the solution at a point x may be expressed in terms of its value at $2^{-n}x$, for any $n > 0$. In the case of the functional equation at hand, the recurrence relation happens to define the Chebyshev polynomials. Although such a particular coincidence is not mandatory in treating general functional equations of the types at hand, we may use this recurrence relation to extend the approximation $v(x)$ from smaller values to large. In particular

$$v(s) := T_{2^n}(v(2^{-n}s)). \tag{4.2.1-5}$$

In particular, if $v(x)$ is known for $0 \le x \le t$, then this formula can be used to define it for those values of s for which $2^{n-1}t \le s \le s^n t$. As the basic interval $[0, t]$ shrinks with t, the accuracy of the approximation increases since our methods are composed of smooth operations. We can evaluate the polynomial in (4.2.1-5) to an arbitrary accuracy (cf. [5]). Thus $v(s)$ can be obtained to any required accuracy by our methodology.

These observations lead us to approximate $y(x)$ for, say, $N = 2$ in domains of the form $X = [0, 2^{-n}]$. The ansatz $v(x) = a + bx^2$ (recall the symmetry) leads to

$$a + bx^2 = 2a^2 + abx^2 + \frac{1}{8}b^2x^4 - 1. \tag{4.2.1-6}$$

The boundary condition itself is translated to X by the recurrence relation itself. In particular, we take $r_0 = 0$, $r_i = \sqrt{(r_{i-1}+1)/2}$, $i = 1,...,n$. Then setting $r := r_n$, we see that

$$y(2^{-n}) = r \tag{4.2.1-7}$$

is an appropriate boundary condition. (Note that for the equation at hand, r is the smallest nonnegative zero of $T_{2^n}(x)$.) (4.2.1-7) gives

$$r = a + 2^{-2n}b := a + u. \tag{4.2.1-8}$$

Referring to Section 3.2.1(iii), we see that relevant roundings that are to be applied to (4.2.1-6) are

$$S_0(x^2) = 2^{-2n-1} \quad \text{and} \quad S_0(x^4) = 3 \times 2^{-4n-3}.$$

Assembling, we get

$$-a - \frac{1}{2}u + 2a^2 + \frac{1}{2}ua + \frac{3}{64}u^2 - 1 = 0,$$

$$u + a - r = 0. \tag{4.2.1-9}$$

Eliminating u $(u = r - a)$ yields a quadratic to be coupled with (4.2.1-8) for solving for b. The resulting system is

$$99a^2 + (-32 + 26r)a - (64 + 32r - 32^2) = 0,$$

$$b = (r - a)2^{2n}.$$

$v(1/2)$ is obtained from $v(2^{-n-1}) = (3a + r)/4$ by using (4.2.1-5). In particular, setting $s_0 = v(2^{-n-1})$ and $s_i = 2s_{i-1}^2 - 1$, $i = 1,...,n$, we have $v(1/2) = s_n$.

We summarize a sample of solution results in Table 4.1.2-1.

n	a	b	$v(2^{-n-1})$	$\left\|\frac{y-v}{y}\right\|$	$v\left(\frac{1}{2}\right)$
0	0.9817286	−0.9817286	0.736295	$4.0 \cdot 10^{-2}$	0.736295
1	0.99860318	−1.1659856	0.9257291	$1.0 \cdot 10^{-2}$	0.71394865
2	0.99990852	−1.2164637	0.9809013	$2.4 \cdot 10^{-3}$	0.70878890
3	0.99999421	−1.2293718	0.99519198	$6.0 \cdot 10^{-4}$	0.70752545
4	0.99999964	−1.2326170	0.9987591	$1.5 \cdot 10^{-4}$	0.70721113
5	0.99999998	−1.2334282	0.99969885	$3.6 \cdot 10^{-5}$	0.70713161

Table 4.1.2-1

From Table 4.1.2-1 we see that with increasing n, $v(1/2)$ converges to $\sqrt{2}/2$ and that $|(y - v)/y|$ diminishes at the rate of $1/4$. However, we expect these properties to persist only if rounding errors are avoided. The rate of convergence may be improved by increasing the degree of approximation N.

4.2.2 The Infinitesimal Case

We confine our attention to the class of nonlinear problems, $\mathscr{L}y = 0$, generated as algebraic combinations in terms of the operations in $(\mathscr{M}; \Omega) = (\mathscr{M}; +, -, \bullet, /, \int)$. We emphasize that this restriction does not confine us to algebraic operators. Indeed, transcendental operators of many sorts are captured by this framework. An example will illustrate this point.

Consider the differential equation

$$\mathscr{L}y = y'' + J_p(y)y' + e^x y + \sqrt{1 + y} \equiv 0 \qquad (4.2.2\text{-}1)$$

supplied with appropriate boundary conditions. Here $J_p(t)$ is the Bessel function of order p defined as an appropriate solution of the differential equation

$$t^2 \tilde{u}'' - 2t\tilde{u}' + (p^2 - t^2)\tilde{u} = 0. \qquad (4.2.2\text{-}2)$$

Setting $u(t) = J_p(y(t)) = \tilde{u}(y(t))$ so that $u'(t) = \tilde{u}'(y)y'$, we find

$$u''(t) = \tilde{u}''(y)y'^2 + \tilde{u}'(y)y''. \qquad (4.2.2\text{-}3)$$

For the coefficient e^x in (4.2.2-1), we have

$$v' - v = 0, \quad v(0) = 1, \qquad (4.2.2\text{-}4)$$

while for the coefficient $w = \sqrt{1 + y}$ we write

$$w^2 - y = 1. \qquad (4.2.2\text{-}5)$$

Combining (4.2.2-2)-(4.2.2-5), we formulate $\mathscr{L}y = 0$ in an equivalent algebraic form:

$$\tilde{\mathscr{L}}(z) = \begin{vmatrix} y''(x) + u(x)y'(x) + v(x)y + w(x) \\ y'y^2u'' - (y'' + 2yy'^2) + (p^2 - y^2)y'^3u \\ v' - v \\ w^2 - y - 1 \end{vmatrix} \equiv 0, \qquad (4.2.2\text{-}6)$$

combined, of course, with all relevant boundary conditions. Clearly, $\tilde{\mathscr{L}}(z)$ composed of algebraic operations in $(\mathscr{M};\Omega)$ defines the extended function $z = (y,u,v)$, the second component of which is the Bessel function.

The differential equations in (4.2.2-6) may be written as systems of first order differential equations. This motivates consideration of a model nonlinear equation of the form

$$0 = \mathscr{L}y := y' - f(x,y), \quad -1 \le x \le 1, \qquad (4.2.2\text{-}7a)$$

with the initial condition

$$y(0) = \eta. \qquad (4.2.2\text{-}7b)$$

This initial-value problem is converted into the integral equation

$$v = \eta + \int_0^x f(x,v(x))dx, \qquad (4.2.2\text{-}8)$$

the latter composed of a compact mapping appropriate to the fixed point theorems of Chapter 2 to which we shall ultimately appeal for validation. (4.2.2-8) in turn is replaced by a Picard-Lindolöff iteration:

$$v_{i+1} = \eta + \int_0^x f(x,v_i(x))dx, \quad i = 0,1,\dots, \qquad (4.2.2\text{-}9)$$

v_0 being given.

The most direct way to illustrate the manner in which to implement (4.2.2-9) in the functoid $(S_N(\mathscr{M});S_N(\Omega)) = (S_N(\mathscr{M}); \boxplus, \boxminus, \boxdot, \boxslash,$ $\text{⨋})$ is to suppose that $f(x,y)$ is composed of its arguments through rational combinations of the five operations $+,-,\bullet,/,\int$.

We indicate this by writing

$$f(x,y) = f(x,y;\Omega).\qquad(4.2.2\text{-}10)$$

Then the Picard-Lindolöff iteration (4.2.2-9) is implemented in $(S_N(\mathcal{M}); S_N(\Omega))$ as follows:

$$v_{i+1}(x) = \eta \boxplus \oint_0^x f(x,v_i(x); S_N(\Omega))dx, \quad i = 0,1,\dots .\qquad(4.2.2\text{-}11)$$

Here the $v_i(x) \in S_N(\mathcal{M})$ (for instance, the $v_i(x)$ are polynomials of degree N). This and other possibilities of computer implementation of equations are discussed in detail in Section 7.1.

The iteration (4.2.2-11) is stopped after k steps, say, by some termination criterion producing an approximation $\tilde{y} = v_k$ of the solution. Of course the solution of (4.2.2-8) is unique. However, in the nonlinear case in general, the final solution to be validated is highly dependent on the choice of the approximation $\tilde{y}$.

For the validation step, we apply the mean-value theorem to (4.2.2-7) and (4.2.2-8) to derive the equation

$$y(x) = \eta + \int_0^x f(x,\tilde{y})dx + \int_0^x \frac{\partial}{\partial y}f(x,s)\,(y - \tilde{y})dx,\qquad(4.2.2\text{-}12)$$

where $s = s(x,y,\tilde{y}) \in y(x) \underline{\cup} \tilde{y}(x)$. Here $\underline{\cup}$ denotes the convex union, so that $y(x) \underline{\cup} \tilde{y}(x)$ is a line segment.

Then recalling the assumption about the form of f (cf. (4.2.2-10)), we use (4.2.2-12) and the operations in $IS_N(\Omega) = \{\,\Diamond\!\!\!\!+, \Diamond\!\!\!\!-, \Diamond\!\!\!\!\times, \Diamond\!\!\!\!/, \oint\,\}$ to define the following function:

$$F(x,\tilde{y},V; IS_N(\Omega)) := \eta \;\Diamond\!\!\!\!+\; \oint_0^x f(x,\tilde{y}, IS_N(\Omega))dx$$

$$(4.2.2\text{-}13)$$

$$\Diamond\!\!\!\!+\; \oint_0^x \frac{\partial}{\partial y}f(x,V(x), IS_N(\Omega)\; \Diamond\!\!\!\!\times\; (V(x)\; \Diamond\!\!\!\!-\; \tilde{y}(x))dx.$$

Now define the following iteration in the interval-functoid $(IS_N(P\mathcal{M}); IS_N(\Omega))$ (cf. (3.2-5)f):

$$V_{i+1}(x) := F(x, \tilde{y}(x), V_i(x), IS_N(\Omega)), \quad i = 0, 1, ..., \qquad (4.2.2\text{-}14)$$

with $V_0 := \tilde{y}$. This iteration is to be conducted until (if ever) an index $i = k$ is reached such that $V_{k+1} \overset{\circ}{\subset} V_k$. Then (as will be shown in Section 7.1 below), we may appeal to Theorem 4 in Chapter 2 to provide the following validation for the solution $y(x)$ of (4.2.2-8): $y(x)$ exists, is unique, $y(x) \in V_{k+1}(x)$ so that the approximation error is less than $d(V_{k+1}(x))$ uniformly for $x \in [-1, 1]$.

examples

For our first example, we note that the sample problem (4.2.1-3) corresponding to the infinitesimal case treated in Section 4.2.1 may also be treated within the framework of the present infinitesimal case methodology. In particular the evaluation of the function $f(x)$ in (4.2.1-3) abbreviated here as

$$y(x) := f(x) := \cos(p(x)) + \gamma, \qquad (4.2.2\text{-}15)$$

may proceed through the solution of the following nonlinear initial value problem:

$$\begin{aligned} y' &= u, \\ p'u' &= -(p')^3 y + p''u. \end{aligned} \qquad (4.2.2\text{-}16a)$$

The initial conditions may be chosen, say, as

$$y(\xi_1) = \cos(p(\xi_i)) + \gamma, \quad i = 1, 2 \qquad (4.2.2\text{-}16b)$$

where $\xi_i \in [-1, 1]$, $i = 1, 2$. These two particular values $y(\xi_i)$ must be found by independent means. Of course the problem (4.2.2-16) is simply a special case of the problem (4.2.2-7), which we have just discussed.

Further examples of nonlinear problems in the infinitesimal case arise through the treatment of linear problems by nonlinear methods. We treat several model problems in this way. First, we provide an approximation in an appropriate functoid and then validation through computation in a corresponding interval-functoid.

Consider then the model linear initial-value problem:

$$y' = y, \quad y(0) = 1. \tag{4.2.2-17}$$

We employ the technique of Padé or Maehly approximation and approximate its solution $y(x)$ by a rational function $v(x) = p(x)/q(x)$. In terms of p and q, (4.2.2-17) becomes

$$qp' - q'p = pq, \quad p(0) = q(0), \quad q'(0) = 1,$$

the last condition being an arbitrary choice of normalization. Employing the ansatz $p = a + bx + cx^2$ and $q = a + x$ and by employing Chebyshev rounding (i.e., Maehly approximation), we find the following equations for a, b, and c:

$$ab - a = a^2,$$
$$2ac + b - b = ab + a + \frac{3}{4}c,$$
$$2c - c = ac + b.$$

Then $b = a + 1$, $c = (a + 1)/(1 - a)$ and $a^3 + 3a^2 - 3/4a - 3/4 = 0$. Thus

$$a = -3.162174,$$
$$b = -2.162174,$$
$$c = -0.5204117.$$

Note that

$$e \sim v(1) = \frac{a + b + c}{a + 1} = 2.7005718,$$

and

$$\left| \frac{e - v(1)}{e} \right| \le 0.0066,$$

a rather good approximation.

Our second example deals with differential equations with a singular coefficient of the highest derivative term. We consider nonlinear methods for this class of problems, a generic member of which is

$$a(x)y'' + b(x)y' + c(x)y = d(x), \quad y(-1) = r, \ y(1) = s. \qquad (4.2.2\text{-}18)$$

We suppose that $a(x)$ has zeros in the domain $X = [-1, 1]$. Since y'' may be infinite at such zeros, the solutions of this problem cannot be expected to be included by polynomials. To circumvent this difficulty we introduce the parametric representation

$$y = \phi(t), \quad x = \psi(t), \quad y' = z = \theta(t),$$

which transforms (4.2.2-18) into

$$
\begin{aligned}
\dot{\phi}(t) &= \theta(t), \\
\dot{\theta}(t) &= -b(\psi(t))\theta(t) - c(\psi(t))\phi + d(\psi(t)), \qquad (4.2.2\text{-}19) \\
\dot{\psi}(t) &= a(\psi(t)),
\end{aligned}
$$

with the following boundary conditions

$$
\begin{aligned}
\phi(\tau) &= r, \qquad \phi(\sigma) = s, \\
\psi(\tau) &= -1, \quad \psi(\sigma) = 1.
\end{aligned}
$$

τ and σ are unknown values of t to be determined, subject to appropriate constraints, as part of the solution.

It is sufficient for our illustrative purposes to replace (4.2.2-19) by our model problem (4.2.2-7), which in parametric form becomes

$$\dot{\phi} = \dot{\psi}\phi, \quad \phi(\tau) = 1, \quad \psi(\tau) = 0. \qquad (4.2.2\text{-}20)$$

We take the ansatz $\phi = a + bt + ct^2$ and $\psi = \alpha + \beta t + \gamma t^2$. The homogeneity of the problem enables us to select a normalization for ψ, which we do by means of the assignments

$$\psi(1) = 1, \quad \psi(-1) = -1. \qquad (4.2.2\text{-}21)$$

These in turn imply that

$$\alpha + \gamma = 0, \quad \beta = 1.$$

Now upon employing the Chebyshev roundings $S_1 x^2 = 1/2$, $S_1 x^3 = 3/4x$, and by our customary methods, we are led to the system of equations

$$\dot{b} = a,$$
$$2c = b + 2\gamma a + \frac{3}{2}\gamma c,$$
$$0 = c + 2\gamma b,$$

additional constraints $\begin{cases} \beta = 1, \\ \\ \alpha + \gamma = 0, \end{cases}$ (4.2.2-22)

boundary conditions $\begin{cases} a + b\tau + c\tau^2 = 1, \\ \\ -\gamma + \tau + \gamma\tau^2 = 0. \end{cases}$

Using the first and third equations to eliminate a and c from the second gives

$$3\gamma^2 - 6\gamma - 1 = 0.$$

The relevant solution for this is $\gamma = 1 - 2/\sqrt{3}$. The last equation can now be solved for τ, whose relevant value is

$$\tau = \frac{-1 + \sqrt{1 + 4\gamma^2}}{2\gamma} = -0.15116547.$$

Using $a = b$ and $c = -2\gamma b$ now gives

$$b = \frac{1}{1 + \tau - 2\gamma\tau^2} = 1.1603544.$$

Then

$$e \sim \phi(1) = a + b + c = 2b(1 - \gamma) = 2.6981990$$

with

$$\left| \frac{\phi(1) - e}{e} \right| \leq 0.0074,$$ (4.2.2-23)

while

$$e^{-1} \sim \phi(-1) = a - b + c = c = -2\gamma b = 0.3590149,$$

$$\left| \frac{\phi(-1) - e^{-1}}{e^{-1}} \right| \leq 0.087.$$

(4.2.2-24)

VALIDATION

Let us return to the problem (4.2.2-20) and compute an inclusion for its solution. There are two unknown functions ϕ and ψ and only one equation. Thus, either ϕ or ψ may be chosen conveniently and the remaining function determined. To use one of our fixed-point theorems, we select ψ and consider the following integral equation derived from (4.2.2-20):

$$\phi = \int_{\tau}^{t} \dot{\psi}\phi \, dt + 1.$$

(4.2.2-25)

ψ and τ are taken as determined by the previous pointwise derivation, viz.,

$$\psi(t) = -\gamma + \beta t + \gamma t^2$$

and

$$\tau = (-1 + \sqrt{1 + 4\gamma^2})/2\gamma,$$

with $\gamma = 1 - 2/\sqrt{3}$ and $\beta = 1$.

Now using the ansatz $\phi = A + Bt + Ct^2$ where A, B, and C are intervals, (4.2.2-25) yields

$$A + Bt + Ct^2 \supset 1 + \int_{\tau}^{t} (\beta + 2\gamma t)(A + Bt + Ct^2) dt. \quad (4.2.2-26)$$

The ensuing computation is simplified if we make the following transformation

$$A = A^* - B^*\tau + C\tau^2,$$
$$B = B^* - 2C\tau,$$

where A^* and B^* are intervals. This transformation and the ansatz gives

$$\phi(t) = A^* + B^*(t - \tau) + C(t - \tau)^2.$$

Likewise setting $\beta^* = \beta + 2\gamma\tau$, we obtain

$$\psi = \beta^* + 2\gamma(t - \tau)^2.$$

Now set $t - \tau = s$, so that $t \in [-1, 1]$ requires that

$$s \in [-1 - \tau, 1-\tau].$$

Denoting the maximum value of s in this interval by λ, we have $\lambda = 1 - \tau \leq 1.15117$. Assembling these changes, the inclusion relation (4.2.2-26) becomes

$$A^* + B^*s + Cs^2 \subset 1 + \int_0^s (\beta^* + 2\gamma s)(A^* + B^*s + Cs^2)ds. \quad (4.2.2-27)$$

Integrating this relation and employing the directed Chebyshev round-ings for the interval $[-1, 1]\lambda$:

$$IS_2s^3 = \frac{3}{4}\lambda^2 s + \frac{1}{4}\lambda^3 \theta, \quad \theta = [-1, 1],$$

$$IS_2s^4 = \lambda^2 s^2 - \frac{1}{4}\lambda^4 \sigma, \quad \sigma = [0, 1],$$

we are led to the following system of inclusion relations for the determi-nation of A^*, B^*, and C:

$$A^* \supset 1 + \frac{\lambda^3 \theta}{12}(\beta^*C + 2\gamma B^*) - \frac{1}{8}\lambda^4 \sigma\gamma C,$$

$$B^* \supset A^*\beta^* + \frac{\lambda^2}{4}(\beta^*C + 2\gamma B^*), \quad (4.2.2-28)$$

$$C \supset \frac{1}{2}(\beta^*B^* + 2A^*\gamma) + \frac{\lambda^2}{2}\gamma C.$$

The solution of (4.2.2-28) is

$$A^* = [0.99315, \quad 0.99316],$$
$$B^* = [1.05181, \quad 1.05182],$$
$$C = [0.32925, \quad 0.41271].$$

With these values,

$$y = \phi(t) \in V(t) := A^* + B^*(t - \tau) + C(t - \tau)^2,$$
$$x = \psi(t) = \alpha^* + \beta^*(t - \tau) + \gamma(t - \tau)^2$$

represents an inclusion of $y(x)$ in parametric form.

To incorporate the constraints $\psi(\pm 1) = \pm 1$, set $s_1 := -1 - \tau$ and $s_2 := 1 - \tau$. Then

$$y(1) \quad = e \quad \in A^* + B^* s_1 + C s_1^2 = [2.6402, \ 2.7509],$$

$$y(-1) = e^{-1} \in A^* + B^* s_2 + C s_2^2 = [0.33757, \ 0.39771].$$

Chapter 5

ITERATIVE RESIDUAL CORRECTION

The method of iterative refinement or iterative residue correction (IRC) is a well-known computational technique for improving the accuracy of an approximation to the solution of equations, especially linear equations [16]. Until recently this method was applied in the context of problems cast in modular number systems such as floating-point representation systems. In this chapter we consider such iterative residue correction methods in function spaces, in particular, in the computational framework of functoids.

In Chapter 3 we showed that functoids and their corresponding roundings have a great similarity to floating-point number structures and their roundings. In the context of functoids novel features and problems for iterative residue correction emerge. Some of these novel features have counterparts in the traditional floating-point context as a deeper review of IRC in that floating-point context reveals. Indeed, we take such a review as our starting point since it will serve to motivate much of our following treatment. We begin in Section 5.1 with a review of the IRC process in a floating-point system. We discuss two arithmetic phenomena associated with IRC. The first is the need for increasing accuracy in the computation of residuals. The second is a description of the propagation of information among the digits in a floating-point system that the IRC process engenders in a floating-point system. It is this latter phenomenon that generates the features of IRC in the framework of a functoid that is central to the ideas in this chapter.

Indeed, following this we turn to two protocols of IRC in a function space in Sections 5.2 and 5.3. In Section 5.2 we consider a model problem corresponding to a Volterra integral equation and an IRC process for it. The IRC process refines the approximation of the function (the solution of the integral equation) in a manner that characterizes a flow of information from low-order basis elements of the function space to high-order basis elements. This process is an analogue of flow of information from left to right among the digits in the floating-point IRC process as described in Section 5.1.

In Section 5.3 a model problem is treated for which the associated IRC process exhibits an analogue of the floating-point feature of carry among the digits. Indeed, this process is characterized by information flow in both directions; low to high and high to low (the carry feature) basis elements of the function space.

The processes of Sections 5.2 and 5.3 are ad hoc, and in Section 5.4 we turn to the development of a formal additive iterative correction process of which both of the former examples are special cases. The formal process are described, and then in Section 5.4.1 its connection to a block relaxation process with steering is established. This process is conducted on the infinite linear system, which is the isomorphic counterpart in terms of a basis of a linear functional equation set in a separable Hilbert space. A convergence criteria for the block process is also developed. Details of how the ad hoc processes of Sections 5.2 and 5.3 correspond to the block process are given in Section 5.4.2 along with examples of application of the convergence criterion.

5.1 ARITHMETIC IMPLICATIONS OF IRC

In this Section we review the IRC process in a floating-point system with particular emphasis on two arithmetic features:
(i) the need for increasing the accuracy of computation of residuals during the process;
(ii) the propagation of information among digits during the process.

The latter feature is described in a context for achieving annihilation of digits in the residuals, and it is this feature that motivates our subsequent treatment of IRC in function spaces.

Statement of the IRC Process
Consider the linear problem

$$\ell x = r \tag{5.1-1}$$

for the unknown vector x.

Let g be an approximate inverse of ℓ arising in some solving process, such as a cycle of SOR or an incomplete Gaussian elimination process. We take

$$\tilde{x} := gr \tag{5.1-2}$$

as an approximation to the solution x of (5.1-1). Since $g\ell - E \neq 0$, in general, we expect that $\tilde{x} \neq x$.

The deviation $\Delta x = x - \tilde{x}$ is sought next, and an equation determining it is obtained directly from (5.1-1):

$$\ell \Delta x = r - \ell \tilde{x}. \tag{5.1-3}$$

The right-hand side of (5.1-3) is the residual of (5.1-1) at the approximation $\tilde{x}$.

Of course (5.1-3) as an equation of the form (5.1-1) is no less difficult to solve. Indeed, the approximate inverse g is used once again, this time with (5.1-3), to produce an approximation $\overset{\sim}{\Delta x}$ to Δx:

$$\overset{\sim}{\Delta x} := g(r - \ell \tilde{x}). \qquad (5.1\text{-}4)$$

With $\overset{\sim}{\Delta x}$ determined, a new and a presumably improved approximation

$$\overset{\approx}{x} := \tilde{x} + \overset{\sim}{\Delta x}$$

is available. $\overset{\sim}{\Delta x}$ may be viewed as a correction to $\tilde{x}$ obtained in terms of the residual of the latter via (5.1-4). Hence this process of passing from $\tilde{x}$ to $\overset{\approx}{x}$ is called a residual correction step (RCS). Iterating such a RCS is called IRC. IRC is succinctly defined by the following recurrence relations:

$$
\begin{aligned}
\text{a)} \quad & \overset{\sim}{\Delta x}_{i+1} := gr_i, \\
\text{b)} \quad & \tilde{x}_{i+1} := \tilde{x}_i + \overset{\sim}{\Delta x}_{i+1}, \\
\text{c)} \quad & r_{i+1} := r_i - \ell \overset{\sim}{\Delta x}_{i+1} \\
& \qquad\; = r_0 - \ell \overset{\sim}{\Delta x}_{i+1},
\end{aligned}
\qquad (5.1\text{-}5)
$$

for $i = 0,1,2,...$ with

$$r_0 := r, \quad \tilde{x}_0 := 0.$$

Here $\tilde{x}_i$ is the ith approximation to the solution x of (5.1-1) produced by this IRC process. $\overset{\sim}{\Delta x}_{i+1}$ is the correction to $\tilde{x}_i$ produced by the residual correction process. $r_i = r - \ell \tilde{x}_i$ is the residual of (5.1-1) at $\tilde{x}_i$. Using this expression for r_i and (5.1-5), we have

$$
\begin{aligned}
\tilde{x}_{i+1} &= \tilde{x}_i + \Delta x_{i+1} \\
&= \tilde{x}_i + gr_i \\
&= \tilde{x}_i + g(r - \ell \tilde{x}_i) \\
&= (E - g\ell)\tilde{x}_i + gr.
\end{aligned}
\qquad (5.1\text{-}6)
$$

Thus IRC is equivalent to the mapping of successive approximations

$\tilde{x}_i \to \tilde{x}_{i+1}$, defined by

$$\tilde{x}_{i+1} := gr + (E - g\ell)\tilde{x}_i, \quad i = 0,1,2,\ldots, \tag{5.1-7}$$

with $\tilde{x}_0 = 0$. Thus the IRC process is convergent if and only if $\sigma(E - g\ell) \leq \kappa < 1$, for some constant κ.

A Practical Problem for IRC Arising in Floating-Point

The representation (5.1-5) for IRC conceals a practical problem that arises when these formulas are actually implemented in floating-point. Of course in that case the relations in (5.1-5) are replaced by rounded versions in which the arithmetic operations are replaced by approximate operations, viz., the floating-point arithmetic operations of a particular computer.

Suppose that our floating-point system employs mantissas of length d. Suppose also that the approximation $\tilde{x}_i$ is in fact good to a certain number of places say $p \leq d$, of its floating-point mantissa representation. Then in order for $\tilde{x}_{i+1}$ in (5.1-5b) to correct $\tilde{x}_i$, Δx_{i+1} itself must be accurately determined in mantissa places beyond those for which $\tilde{x}_i$ is known, i.e., beyond the pth place. But Δx_{i+1} is determined from (5.1-5a). Thus the calculation of gr_i in (5.1-5a) must be made so that it is accurate beyond the pth place. Next we observe that r_i itself is determined from (5.1-5c) (with i replaced by $i - 1$). Thus r_i as it depends on r_{i-1} and x_i must be calculated to more than p places of accuracy. (The actual number of k of places beyong the pth needed depends on the condition of ℓ.)

These accuracy requirements are typically achieved only in a preliminary way, and then even by happenstance in typical implementations of IRC on a computer. This is the case since p, the number of accurately known digits of $\tilde{x}_i$, starts out as a small number, much less than the mantissa length d. Then the residue calculation of (5.1-5c) when carried out to d places typically produces a value of r_i accurate to more than p places. The same is true for (5.1-5a) and (5.1-5b). Thus in practice the IRC process starts out well, that is, it produces a sequence of approximations $\{\tilde{x}_i\}$ that demonstrates steady improvement in the numerical sense (i.e., by acquiring more and more invariant initial digits in its representations). As may be expected, this state of affairs quickly dissipates as p increases. Indeed, p may never become particularly significant.

At this point in the IRC process, continued effectiveness may be injected into it by continuing the calculation in (5.1-5) in double precision (i.e., by replacing d by $2d$) or even in extended precision (i.e., by replacing d by $4d$) etc. This is the common practice, and for many problems this serves to supply the desired accuracy in the approximation of the solution of the equation (5.1-1) being solved. Of course there are ramified and ill-conditioned problems for which this stepwise increase of precision process does not work well at all.

All of this points out the dependence and sensitivity of the IRC process on the quality of the arithmetic that is employed in actual computer (i.e., floating-point) implementation. In fact, rather complete and effective solutions of this arithmetic aspect of IRC have been found by employing optimal computer arithmetic and, in particular, optimally accurate inner products [9], [11], [14].

Propagation of Information among Digits in IRC

The more compact form (5.1-7) of the IRC process enables us to interpret these arithmetic features in terms of the approximate inverse operator g. Indeed, computer implementation of (5.1-7) may be regarded as replacement of g therein by a new approximate inverse describing both the action of g and the roundings employed in evaluation of (5.1-7).

In fact, the approximate inverse g may be changed from time to time during the IRC process in order to achieve greater accuracy, equivalently a more rapid convergence, or, for example, to achieve a reduction in work performed, etc. Thus g may be replaced by g_i, an iteration-dependent approximate inverse. From a formal point of view a sufficient condition for such variations in g to be permissible is that $\sigma(E - g_i \ell) \leq \kappa < 1$, for almost all $i \geq 0$.

We turn now to a particular example of variation in the choice of g that demonstrates influence on the movement of information between digits in a floating-point representation of the IRC process. This example will also anticipate certain analogous function-space features to follow wherein information propagates between expansion coefficients in the functoid.

Let $x \in \mathbb{R}$ have the floating-point representation

$$x = \sum_{i=1}^{\infty} \zeta_i 10^{-i} \cdot 10^e,$$

where e is some exponent.

Let $S_N^p : \mathbb{R} \to S_N(\mathbb{R})$ be a rounding operator:

$$S_N^p x = S_N^p \sum_{i=1}^{\infty} \zeta_i 10^{-i} \cdot 10^e$$

$$= \sum_{j=p}^{p+N-1} \zeta_j 10^{-j} \cdot 10^e$$

$$= \sum_{k=1}^{N} \zeta_{p+k-1} 10^{-k} \cdot 10^{e+p-1}.$$

Thus S_N^p cuts a word out of x of (mantissa) length N beginning with the pth position, as the following schematic shows:

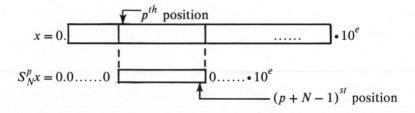

S_N^p as a rounding operator appears to be unusual since it does not take the leading part of x, but rather some slice of x "somewhere in its middle". While such a rounding is unusual in the case of floating-point numbers, it plays an essential role in functoids, as we shall see below.

Let R be an approximate inverse to ℓ, and let

$$g_i = S_N^{q_i} R\ S_N^{p_i}, \tag{5.1-8}$$

where q_i and p_i are integers depending on i. In a sense g_i is an "implemented" version of R since it is preceded and succeeded by rounding operators.

Now let us represent the IRC process by the single equation

$$r_{i+1} := r_i - \ell g r_i \tag{5.1-9}$$

obtained by combining (5.1-5a) and (5.1-5c). Now combining (5.1-8) and (5.1-9) we get

$$r_{i+1} := r_i - \ell(S_N^{q_i} R S_N^{p_i}) r_i. \tag{5.1-10}$$

We may view the residual r_i as an indicatrix of error size at step i in the solution process. Then a strategy for error reduction is to choose p_i and q_i so that some of the leading digits of r_i are annihilated when r_{i+1} is computed from (5.1-10). To do this, we set $p_i = 1$ so that $S_N^{p_i}$ deletes no leading digit of r_i. Similarly we set $q_i = 1$ so that a maximum amount of leading numerical information of $R S_N^{p_i} r_i$ is likewise retained. (With these choices $p_i = q_i = 1$, the heretofore unusual quality of the roundings $S_N^{p_i}$ and $S_N^{q_i}$ has disappeared, at least for now. Indeed, these roundings now cut out N leading digits.)

Now, since $g = S_N R S_N$ is an approximate inverse of ℓ, we suppose that some of the leading digits of $g r_i$ and of $\ell^{-1} r_i$ coincide. Indeed, suppose that

$$r_i = \overset{\circ}{r}_i + \hat{r}_i, \tag{5.1-11}$$

where $\overset{\circ}{r}_i$ are those leading digits of r_i for which this coincidence occurs, i.e., for which

$$g\overset{\circ}{r}_i = \ell^{-1}\overset{\circ}{r}_i.$$

Then

$$\ell g \mathring{r}_i = \mathring{r}_i. \tag{5.1-12}$$

We may view (5.1-12) as an equation for determining $\mathring{r}_i$, that is, an equation for the determination of which, if any, of the leading digits of r_i cancel in the determination of r_{i+1} in (5.1-10). Now employing linearity of ℓ and g and combining (5.1-10) and (5.1-11), we get

$$r_i - \ell g r_i = \mathring{r}_i + \hat{r}_i - \ell g(\mathring{r}_i + \hat{r}_i)$$

$$= \mathring{r}_i + \hat{r}_i - \ell g \mathring{r}_i - \ell g \hat{r}_i \tag{5.1-13}$$

$$= \hat{r}_i - \ell g \hat{r}_i,$$

the last following from (5.1-12). This is the value of r_{i+1}.

Let $\bar{r}_i$ and $\bar{r}_{i+1}$ be fixed-point representations of r_i and r_{i+1}, respectively, in a base β. From (5.1-9) and the property $\rho(E - \ell g) \le x < 1$, we have that $|r_{i+1}| < |r_i|$. Thus for any fixed integer $t > 1$, there exists a base $\beta = \beta(x,t) < 1$ such that $\bar{r}_{i+1}$ has t more leading zero digits (of base β) than does $\bar{r}_i$.

Now suppose $\mathring{r}_i$ contains $t \ge 2$ digits. The term $\hat{r}_i$ starts with the $(t + 1)^{st}$ digit. In principle the remaining term $\ell g \hat{r}_i$ could contain digits anywhere and, in particular, among the first t, thereby undoing the cancellation of digits represented by (5.1-12). We shall show that the two terms $\mathring{r}_i$ and $\ell g \hat{r}_i$ overlap in at most one place. That is, $\ell g \hat{r}_i$ has at least $t - 1$ leading zeros. Thus the cancellation of digits represented by (5.1-12) is at most dissipated by one digit in (5.1-13), that is, in the computation of r_{i+1}.

To see this, let

$$\mathring{r}_i = \mathring{\rho}_i \cdot \beta^e \tag{5.1-14}$$

be a normalized floating-point representation for $\mathring{r}_i$ with mantissa $\mathring{\rho}_i$.

Since as we have noted $\hat{r}_i$ starts in the tth position relative to $\overset{\circ}{r}_i$, we have

$$\hat{r}_i = \overset{\wedge}{\rho}_i \bullet \beta^{e-t'}.$$

In fact, $\overset{\wedge}{\rho}_i$ may have leading zeros. So we take

$$\hat{r}_i = \overset{\wedge}{\rho}_i \bullet \beta^{e-t'} \tag{5.1-15}$$

to be the normalized floating-point representation for $\overset{\wedge}{\rho}_i$ increasing t to t' if necessary, but retaining, for convenience, the symbol $\overset{\wedge}{\rho}_i$ for the mantissa. Now the use of the radix β implies

$$\overset{\circ}{\rho}_i \bullet \beta^{-1} \leq \overset{\wedge}{\rho}_i \leq \overset{\circ}{\rho}_i \bullet \beta. \tag{5.1-16}$$

Since $\sigma(E - \ell g) < 1$, then ℓg is near the identity. Then we suppose that ℓg is a positive operator and may be applied to this inequality without changing its sense. Next recalling that ℓg is linear and applying it to (5.1-15) gives

$$\ell g \hat{r}_i = (\ell g \overset{\wedge}{\rho}_i) \beta^{e-t'}.$$

Now combine this with the result of applying ℓg to (5.1-16). We get

$$\ell g \hat{r}_i \leq \max (\ell g \overset{\circ}{\rho}_i \bullet \beta^{-1},\ \ell g \overset{\circ}{\rho}_i \bullet \beta). \tag{5.1-17}$$

Here we use the presumed positivity property of ℓg.

Combining (5.1-12) and (5.1-14), we have

$$\overset{\circ}{\rho}_i \bullet \beta^e = \overset{\circ}{r}_i = \ell g \overset{\circ}{r}_i = (\ell g \overset{\circ}{\rho}_i) \beta^e, \tag{5.1-18}$$

so that

$$\overset{\circ}{\rho}_i = \ell g \overset{\circ}{\rho}_i. \tag{5.1-19}$$

Inserting (5.1-19) into (5.1-17), we have

$$\ell g \hat{r}_i \leq \max \, (\overset{\circ}{\rho}_i \cdot \beta^{-1}, \, \overset{\circ}{\rho}_i \cdot \beta) \cdot \beta^{e-t'}$$

$$= \overset{\circ}{\rho}_i \cdot \beta^{e-t'+1} \qquad\qquad (5.1\text{-}20)$$

$$= \overset{\circ}{r}_i \cdot \beta^{-t'+1},$$

the last following from (5.1-14). Thus we have shown, as claimed, that $\ell g \hat{r}_i$ has at least t fewer leading digits than does $\overset{\circ}{r}_i$ with an exception of at most one place.

The process of cancellation of leading digits that we have just described is associated with the roundings S_N^p for $p = 1$. However, other values of p are in principle equally useful, the object being to annihilate blocks of digits of length N in the residual. Of course, roundings and carries can nibble away at the block, as we have just seen. What we should want is to find a strategy of annihilation of such blocks of digits, perhaps passing back and forth, as in a relaxation process, until as much of the residual (i.e., as many of its digits) as necessary are annihilated. It is relatively unusual to do this in a floating-point computation except for the leading block, as we have seen. In principle this is no reason why we should adhere to this leading-block point of view if, for example, more rapid convergence of the IRC process would accrue by choosing some other policy of block annihilation. In fact, variations of this idea are pursued in our approach to IRC in function spaces to follow. Digits are replaced by their function space analogues: coefficients in a basis expansion (cf. Table 3.1) and blocks of digits by blocks of such coefficients. We turn now to the first of our function space IRC processes.

5.2 IRC FOR INITIAL VALUE PROBLEMS AND VOLTERRA INTEGRAL EQUATIONS

We describe IRC in this case through the treatment of the model problem

$$0 = \mathcal{L}y = 1 + \int_0^x y - y =: 1 + \ell y. \tag{5.2-1}$$

We seek an approximation v of the solution y in $S_N(\mathcal{M})$ using the MPB and Taylor rounding $S_N \colon \mathcal{M}_N \to S_N(\mathcal{M})$ on the domain $X = [-1, 1]$. That is, with the ansatz $v = \sum_{i=0} a_i x^i$, we solve

$$S_N(\mathcal{L}v) = 0 \tag{5.2-2}$$

in the customary manner (cf. Chapter 4). Having determined $v \in S_N(\mathcal{M})$, we seek a correction Δy:

$$y = v + \Delta y. \tag{5.2-3}$$

Inserting (5.2-3) into (5.2-1) gives the *residue correction equation*

$$\ell \Delta y \equiv -\mathcal{L}v. \tag{5.2-4}$$

In turn the correction Δy is approximated by Δv, the latter determined by the rounded equation

$$S_N(\ell \Delta v) \equiv S_N(-\mathcal{L}v). \tag{5.2-5}$$

Now take $N = 2$ so that $v = a + bx + cx^2$. Then

$$S_2(\mathcal{L}v) = S_2\left(1 - a + (a - b)x + (\tfrac{1}{2}b - c)x^2 + \tfrac{1}{3}cx^3 \right)$$

$$= 1 - a + (a - b)x + \left(\tfrac{1}{2}b - c \right)x^2.$$

a, b, and c are determined by inserting this into (5.2-2). We find $v = 1 + x + \tfrac{1}{2}x^2$ and $\mathcal{L}v = \tfrac{1}{3}cx^3 = \tfrac{1}{6}x^3$.
Next (5.2-5) becomes

$$S_2(\ell \Delta v) = S_2\left(-\tfrac{1}{6}x^3 \right). \tag{5.2-6}$$

Note that the rounding here and the ansatz for Δv are not yet specified. As an ansatz we take the *scaled* or *floating polynomial*

$$\Delta v := (a + bx + cx^2)x^m \tag{5.2-7}$$

with the parameter m to be determined. As for the rounding, we note that Taylor rounding applied in the sense of Section 3.1.2(i) would give $S_2(x^3) = 0$. This application of the Taylor rounding is inappropriate since it would result in a null correction. In fact, this application of the Taylor rounding reinjects the residue $-\mathscr{L}v$ into the manifold $S_2(\mathscr{M})$ where we have already solved (determined the approximation for) our problem. We are obliged to interpret $-\frac{1}{6}x^3$ (the residue) as a scaling (shifting). Thus we apply S_2 in the sense of relative roundings in Section $_\varrho$3.1.2(iv), from which we recall that for a polynomial $p(x) = \sum\limits_{i=0} p_i x^i$, we have for $p = m$ that

$$S_2^m(p(x)x^m) = (S_2 p(x))x^m.$$

Now using (5.2-7), we have

$$\ell(\Delta v) = \left(-a + (\frac{a}{m+1} - b)x + (\frac{b}{m+2} - c)x^2 + \frac{a}{m+3}x^3 \right)x^m.$$

Applying S_2^m to this gives

$$S_2^m(\ell(\Delta v)) \equiv \left(-a + (\frac{a}{m+1} - b)x + (\frac{b}{m+2} - c)x^2 \right)x^m.$$

Now using (5.2-5), we have

$$\left(-a + (\frac{a}{m+1} - b)x + (\frac{b}{m+2} - c)x^2 \right)x^m \equiv S_2^m\left(-\frac{1}{6}x^3 \right). \quad (5.2\text{-}8)$$

If for the shifting factor we take $m = 3$, and noting then that $S_2^3\left(-\frac{1}{6}x^3 \right) = -\frac{1}{6}x^3$, (5.2-8) yields

$$-a + \left(\frac{a}{m+1} - b \right)x + \left(\frac{b}{m+2} - c \right)x^2 = -\frac{1}{6}.$$

Thus, $a = \frac{1}{6}$, $b = \frac{1}{24}$, $c = \frac{1}{120}$. For the resulting new approximation we have

$$v_1 := v + \Delta v = 1 + x + \frac{1}{2}x^2 + \frac{1}{6}x^3 + \frac{1}{24}x^4 + \frac{1}{120}x^5, \quad (5.2\text{-}9)$$

a marked improvement over v. Clearly, IRC with relative Taylor rounding in the MPB produces the Taylor expansion of the solution y of (5.2-1). This feature holds for a wide class of linear initial-value prob-

lems and Volterra integral equations. Indeed, the IRC methodology
exhibited for the model problem carries over directly to the more general linear initial-value problem:

$$\mathscr{L}y = \ell y - r = 0, \quad x \in X = [-1, 1]. \tag{5.2-10}$$

We now show this for problems for which
(a)   constraints and initial conditions are defined at the single point
$x = 0$;
(b)   all integrals are of the form $\int_0^t \cdot \, dt$.

The roundings will be taken from among the relative roundings S_N^p
corresponding to a power basis (cf. Section 3.1.2(iv)). The problem
class is to be so restricted that all infinite sums that occur in what
follows are presumed to be convergent series, asymptotic series, etc. In
fact, since we work with finite subspaces, the series are for the most
part finite series.

Let

$$v = w_i + v_{i+1} + v_{i+2} + \cdots, \tag{5.2-11}$$

where $w_i := \sum_{j=0}^{i} v_j$, be the result of successively approximating $y \in \mathscr{M}$
where v_i is defined by the ith residue correction step. In particular, we
take the ansatz

$$v_i := x^{m_i} \sum_{j=0}^{N} a_{ij} x^j := x^{m_i} \bar{v}_i, \quad i = 0, 1, \dots \; . \tag{5.2-12}$$

With this ansatz we may also write

$$w_i = \sum_{k=0}^{m_i + N} c_{ik} x^k,$$

where the c_{ik} depend on the a_{ij}. The choice of the m_i is problem dependent and is to be made strategically. Corresponding to the ansatz
(5.2-12) for the v_i, we choose $m_i = iN' = i(N + 1)$, $i = 0, 1, \dots$. In this
manner the successive corrections v_i, being on the scale m_i, are effected
on that scale (cf. (5.2-17) and (5.2-21) below).

Beginning with $i = 0$ and setting $p = m_0$, we have the equation

$$S_N^{m_0}(\mathscr{L}v_0) = 0, \tag{5.2-13}$$

which we solve as described in Section 4.1.2. Recall that with $m_0 = 0$, $S_N^{m_0} = S_N$, the Taylor rounding. Using (5.2-12) in (5.2-13) we have

$$
\begin{aligned}
0 &= S_N \mathscr{L} \sum_{j=0}^{N} a_{0j} x^j \\
&:= S_N \sum_{j=0}^{\infty} b_{1j} x^j \\
&= S_N \sum_{j=0}^{N} b_{1j} x^j + S_N \sum_{j=N+1}^{\infty} b_{1j} x^j \\
&= \sum_{j=0}^{N} b_{1j} x^j,
\end{aligned}
\tag{5.2-14}
$$

since the sum $\sum\limits_{j=0}^{N}$ here is invariant under S_N while the sum $\sum\limits_{j=N+1}^{\infty}$ is annihilated by S_N.

For $i = 1$, the first residual equation reads

$$S_N^{m_1}(\ell v_1) = -S_N^{m_1} \mathscr{L} v_0. \tag{5.2-15}$$

For the left member here and for our model problem, we have

$$S_N^{m_1} \ell v_1 := S_N^{m_1} \ell x^{m_1} \bar{v}_1 := S_N^{m_1} x^{m_1} \widetilde{\ell} \bar{v}_1 = x^{m_1} S_N \widetilde{\ell} \bar{v}_1. \tag{5.2-16}$$

Recall that the first isomorphic shifting factor is $m_1 = N + 1$. For the right member of (5.2-15) we have (cf. (5.2-14))

$$
\begin{aligned}
S_N^{m_1} \mathscr{L} v_0 &= S_N^{m_1} \mathscr{L} \sum_{j=0}^{N} a_{0j} x^j \\
&:= S_N^{m_1} \sum_{j=0}^{N} b_{1j} x^j + S_N^{m_1} \sum_{j=N+1}^{\infty} b_{1j} x^j \\
&= 0 + x^{N+1} S_N \sum_{j=0}^{\infty} b_{1,j+N+1} x^j \\
&= x^{N+1} S_N \sum_{j=0}^{\infty} b_{1,j+N+1} x^j.
\end{aligned}
\tag{5.2-17}
$$

Now, upon inserting (5.2-16) and (5.2-17) into (5.2-15), we get

$$x^{m_1} S_N \ell \tilde{\bar{v}}_1 = - x^{N+1} \sum_{j=0}^{N} b_{1,j+N+1} x^j.$$

Since $m_1 = N + 1$, we get

$$S_N \ell \tilde{\bar{v}}_1 = - \sum_{j=0}^{N} b_{1,j+N+1} x^j, \qquad (5.2\text{-}18)$$

an equation for the determination of the coefficients of v_1. Note that if in (5.2-17) S_N were not specifically the Taylor rounding, the term $\sum_{j=0}^{N} b_{1,j+N+1} x^j$ in the last member would be replaced by $\sum_{j=0}^{N} \tilde{b}_{1j} x^j$ where the coefficients $\tilde{b}_{1j}$ depend on the coefficients b_{ij}, $j = 0, 1, \dots$.

Proceeding now to the ith step, the ith residual equation reads

$$S_N^{m_i} \ell v_i = - S_N^{m_i} \mathscr{L} w_{i-1}. \qquad (5.2\text{-}19)$$

For the left member we have (cf. (5.2-16))

$$S_N^{m_i} \ell v_i := S_N^{m_i} \ell x^{m_i} \bar{v}_i := x^{m_i} S_N \ell \tilde{\bar{v}}_i, \qquad (5.2\text{-}20)$$

where $m_i (= iN')$ is the ith isomorphic shifting factor. For the right member we have

$$
\begin{aligned}
S_N^{m_i} \mathscr{L} w_{i-1} &= S_N^{m_i} \mathscr{L} \sum_{j=0}^{N+m_{i-1}} c_{i-1,j} x^j \\
&:= S_N^{m_i} \sum_{j=0}^{iN'-1} b_{ij} x^j + S_N^{m_i} \sum_{j=iN'}^{\infty} b_{ij} x^j \\
&= 0 + x^{iN'} S_N \sum_{j=0}^{\infty} b_{i,j+iN'} x^j \\
&= x^{iN'} \sum_{j=0}^{N} b_{i,j+iN'} x^j.
\end{aligned}
\qquad (5.2\text{-}21)
$$

Then combining (5.2-19)-(5.2-21) gives

$$x^{m_i} S_N \ell \tilde{\bar{v}}_i = - x^{iN'} \sum_{j=0}^{N} b_{i,j+iN'} x^j.$$

Hence

$$\widetilde{S_N \ell} \bar{v}_i = - \sum_{j=0}^{N} b_{i,j+iN'} x^j, \tag{5.2-22}$$

an equation for the determination of the coefficients of v_i.

validation

Suppose that $n - 1$ steps of residual correction have been made, result-ing in the approximation w_{n-1} to the solution of our problem. As a final step we employ the directed relative rounding IS_N to obtain an inclusion in $IS_N(\mathcal{M})$ of the solution we seek, where (cf. Section 3.2.2(iv))

$$IS_N(\mathcal{M}) = \{y \mid y = IS_N^p z, z \in \mathcal{M}, p \in \mathbf{N} \cup \{0\}\}.$$

The procedure for obtaining the inclusion is analogous to one used in Chapter 4. Indeed, to obtain an inclusion of the residue we first calcu-late $IS_N^p \mathcal{L} w_n$ (with $p = nN'$) using (3.2.2-21).

$$IS_N^p \mathcal{L} w_n = IS_N^p \mathcal{L} \sum_{j=0}^{N+m_{n-1}} c_{n-1,j} x^j$$

$$= IS_N^p \sum_{j=0}^{nN'-1} b_{nj} x^j + x^{nN'} IS_N \sum_{j=0}^{\infty} b_{n,j+nN'} x^j$$

$$= \sum_{j=0}^{nN'-1} b_{nj} x^j + x^{nN'} \sum_{j=0}^{\infty} b_{i,j+nN'} IS_N x^j.$$

Hence

$$IS_N^p \mathcal{L} w_{n-1} = \sum_{j=0}^{nN'-1} b_{nj} x^j + x^{nN'} \left(\sum_{\substack{j=0 \\ j \text{ even}}}^{\infty} b_{nj}[0, 1] + \sum_{\substack{j=0 \\ j \text{ odd}}}^{\infty} b_{nj}[-1, 1] \right),$$

referring to the subsection on directed Taylor rounding in Section 3.2.2. The sums in this derivation are, in fact, finite since an algebraic operator defined on $(\mathcal{M}; +, -, *, \int)$ takes a polynomial into a polynomial.

Now let ℓy be of the form

$$\ell y := y - \ell y,$$

where ℓ is a compact continuous operator. Let $Y_0 = -IS_N^p \mathscr{L} w_{n-1}$ and perform the following iteration process (cf. (2.1-2)ff):

$$\underline{\text{repeat}} \quad Y_{i+1} := Y_0 + \ell Y_i$$

$$(5.2\text{-}23)$$

$$\underline{\text{until}} \quad Y_{i+1} \subseteq Y_i.$$

Suppose this iteration halts. Then Theorem 1 in Section 2.2.1 implies that an inclusion is obtained. In particular, the last residue correction Δy of y, viz.,

$$y := v_0 + \ldots + v_{n-1} + \Delta y \qquad (5.2\text{-}24)$$

is contained in $Y_{i+1} := Y$. Then the inclusion sought is obtained as well, viz.,

$$y(x) \in v_0(x) + \ldots + v_{n-1}(x) + Y(x). \qquad (5.2\text{-}25)$$

Note that the term Y_0 in (5.2-23) contains the sum of all terms that were formally neglected in the preceding correction steps (cf. the comment following (5.2-21)). Thus, the inclusion process is exact. Moreover, the quality of the inclusion produced will depend on the size of the sum of those neglected terms which are retrieved here.

5.3 ITERATIVE RESIDUAL CORRECTION WITH CARRY

We have already noted the analogy of floating-point numbers with functions represented by expansions in terms of basis functions. This analogy generates the view of IRC as a process of movement of information from left to right in the basis function representation. Arithmetic operations with floating-point numbers, such as division, are frequently implemented by algorithms that develop the mantissa of the result from left to right. On the other hand, computer arithmetic contains the well-known feature of carry, which, say for addition of two floating-point numbers, may be viewed as corresponding to a flow of information from right to left. In fact, arithmetic algorithms frequently contain both forward and backward flows of numerical information. In this section we develop an IRC technique with this feature. It comprises steps of both

> *forward residue correction (fwd RC)*

and

> *backward residue correction (bwd RC)*,

and it is referred to simply as IRC *with carry*.

We develop this method in the context of an example. Namely, we seek to approximate the solution $y(x) = e^x/(e + 1)$ of the boundary-value problem

$$y' = y,$$

$$y(0) + y(1) = 1.$$

(5.3-1)

(5.3-1) is transformed into the integral equation

$$0 = \mathcal{L}y \equiv -r + \ell y \equiv \frac{1}{2} - \frac{1}{2}\int_0^1 y\,dt + \int_0^x y\,dt - y, \quad (5.3-2)$$

with

$$\ell y :\equiv \textit{k}y - y :\equiv -\frac{1}{2}\int_0^1 y\,dt + \int_0^x y\,dt - y \qquad (5.3-3)$$

and $r := -\dfrac{1}{2}$. We shall subsequently make use of the corresponding explicit iteration formula

$$y = -r + \textit{k}y. \qquad (5.3-4)$$

We approximate the solution y of (5.3-2) in $S_N(\mathcal{M})$ on the domain $X = [-1, 1]$ using MPB and a relative Taylor rounding $S_N^p: \mathcal{M} \to S_N(\mathcal{M})$ (cf. section 3.1.2(iv)). We shall systematically call the first approximation v_0, which may be viewed as the zeroth residue correction corresponding to a starting approximation of zero. We begin as in Section 5.2. Then proceeding with the ansatz

$$v_0(x) = (a + bx + cx^2)x^{m_0}, \qquad (5.3-5)$$

we are led to solve

$$S_2^p(\mathscr{L}v_0) = 0. \tag{5.3-6}$$

Setting $p = m_0 = 0$, this gives

$$0 \equiv S_2(\frac{1}{2} - a - \frac{1}{2}(a + \frac{1}{2}b + \frac{1}{3}c) + (a - b)x + (\frac{1}{2}b - c)x^2 + \frac{1}{3}cx^3)$$

$$= \frac{1}{2} - \frac{3}{2}a - \frac{1}{4}b - \frac{1}{6}c + (a - b)x + (\frac{1}{2}b - c)x^2 = 0.$$

For later purposes, we set $\alpha_0(m_0) = \alpha_0(0) = -\frac{1}{2}a - \frac{1}{4}b - \frac{1}{6}c$ and $p_0(x) = -a + (a - b)x + (\frac{1}{2}b - c)x^2$, and we write this equation as

$$0 \equiv S_2(\frac{1}{2} + \alpha_0(m_0) + p_0(x)x^{m_0} + \frac{1}{3}cx^3)$$

$$= \frac{1}{2} + \alpha_0(0) + p_0(x) + S_2(\frac{1}{3}cx^3).$$

We also set $a = a_0$, $b = b_0$, and $c = c_0$.

This leads to the determination $a_0 = b_0 = 6/22$ and $c_0 = 3/22$, so that

$$v_0(x) = \frac{6}{22} + \frac{6}{22}x + \frac{3}{22}x^2. \tag{5.3-7}$$

The quality of this approximation can be surmised from the Table 5.3-1.

x	$v_0(x)$	$y(x)$	$\left|\dfrac{y - v_0}{y}\right|$
0	0.27$\overline{27}$	0.2689	0.014
1	0.68$\overline{18}$	0.7311	0.067

Table 5.3-1

To improve the approximation, we write $y = v_0 + \Delta y$ and proceed to a first residual correction, the equation for which is

$$\ell\Delta y + \mathscr{L}v_0 = 0. \tag{5.3-8}$$

The solution Δy is approximated by v_1, defined by

$$S_2^p(\ell v_1 + \mathscr{L}v_0) \equiv 0, \tag{5.3-9}$$

where p is to be determined. $\mathscr{L}v_0 = \frac{1}{3}c_0 x^3 = \frac{1}{22}x^3$. As in (5.2-7) we employ the shifting factor ansatz

$$v_1 = (a_1 + b_1 x + c_1 x^2)x^{m_1}.$$

We find

$$\ell v_1 = -\frac{1}{2}\left(\frac{a_1}{m_1+1} + \frac{b_1}{m_1+2} + \frac{c_1}{m_1+3}\right)$$

$$+ \left(-a_1 + \left(\frac{a_1}{m_1+1} - b_1\right)x + \left(\frac{b_1}{m_1+2} - c_1\right)x^2 + \frac{c_1}{m_1+3}x^3\right)x^{m_1}$$

$$:= \alpha_1(m_1) + p_1(x)x^{m_1} + \frac{c_1}{m_1+3}x^{m_1+3}.$$

Here $p_1 \in \mathbb{R}^2[x]$ with

$$p_1(x) = -a_1 + \left(\frac{a_1}{m_1+1} - b_1\right)x + \left(\frac{b_1}{m_1+2} - c_1\right)x^2$$

and

$$\alpha_1(m_1) = -\frac{1}{2}\left(\frac{a_1}{m_1+1} + \frac{b_1}{m_1+2} + \frac{c_1}{m_1+3}\right).$$

The condition (5.3-9) now yields

$$S_2^p\left(\alpha_1(m_1) + p_1(x)x^{m_1} + \frac{c_1}{m_1+3}x^{m_1+3} + \frac{1}{22}x^3\right) = 0. \tag{5.3-10}$$

Compare this to the situation in Section 5.2 (cf. (5.2-6)-(5.2-8)). We may set $p = m_1 = 3$, thereby accommodating the forcing term $\frac{1}{22}x^3$ obtained from the previous step. S_2^3 annihilates $\alpha_1(m_1)$ and x^{m_1+3}, and so we are left with

$$S_2^3\left((p_1(x) + \frac{1}{22})x^3\right) = x^3 S_2\left(p_1(x) + \frac{1}{22}\right). \tag{5.3-11}$$

The corresponding equation

$$x^3\left(\frac{1}{22} - a_1 + (\frac{a_1}{4} - b_1) + (\frac{b_1}{5} - c_1)x^2\right) \equiv 0$$

gives $a_1 = \frac{1}{22}$, $b_1 = \frac{1}{88}$, $c_1 = \frac{1}{440}$. For the annihilated term, we find

$$\alpha_1(3) = -\frac{67}{5280}. \tag{5.3-12}$$

Summarizing, we have computed the correction

$$v_1(x) = \left(\frac{1}{22} + \frac{1}{88}x + \frac{1}{440}x^2\right)x^3,$$

and the corrected approximation now reads

$$v(x) = v_0 + v_1$$

$$= \frac{1}{440}\left(120 + 120x + 60x^2 + 20x^3 + 5x^4 + x^5\right). \tag{5.3-13}$$

Since the construction annihilates the term $\alpha_1(3)$ in the residue correction equation (5.3-10), we cannot expect the approximation (5.3-13) to be a good improvement of $v_0(x)$ uniformly on X. Since $v(0) = v_0(0) + v_1(0) = v_0(0)$, no improvement is achieved at $x = 0$ and little in a neighborhood of $x = 0$. However, in contrast to Table 5.3-1, we now have

$$v(1) = 0.7409091 \text{ and } \left|\frac{y(1) - v(1)}{v(1)}\right| \le 0.014.$$

Thus with respect to relative error, the approximation is tending toward uniformity. We may interpret this heuristically by saying that additional information has to be injected into the approximation. To proceed further, we must invoke a new type of residue correction, which accommodates lower basis residue terms such as the annihilated term $\alpha_1 x^0$. Recalling our discussion at the beginning of this subsection, we shall invoke the process of bwd RC, which, in correspondence to the carry of floating-point arithmetic, will transmit information from coefficients of higher-basis elements toward lower-basis elements. This new method is schematized in Fig. 5.3-2.

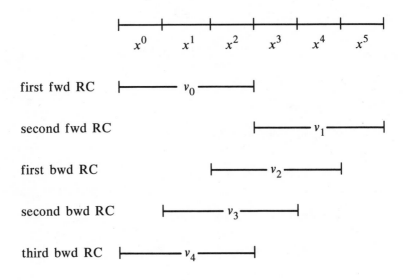

Figure 5.3-2: Schema of IRC with carry

With an ansatz of the form (5.2-7) this diagram represents the choices $m = 0,3,2,1,0$ for the first, second, and third bwd RC steps, respectively. The way in which monomials constituting the backward corrections overlap the monomials constituting v_0 and v_1 schematizes the leftward flow of information that v_2, v_3 and v_4 represent. The steps of residual correction require the evaluation of ℓv_i and of $\mathscr{L}(v_0 + v_1 + ...)$, and we first carry out a side calculation to help us with these evaluations. Set $w(x) = \sum_{i=q}^{p} w_i x^i =: m(w)x^q$, so that $m(w)$ is the mantissa of $w(x)$ and q is its scaling factor (cf. Section 3.2.1(iv)). Then (5.3-2) and (5.3-3) give

$$\ell w = -\frac{1}{2} \sum_{i=q}^{p} \frac{w_i}{i+1} + \sum_{i=q}^{p} \frac{w_i}{i+1} x^{i+1} - \sum_{i=q}^{p} w_i x^i$$

$$(5.3\text{-}14)$$

$$= -\frac{1}{2} \sum_{i=q}^{p} \frac{w_i}{i+1} - w_q + \sum_{i=q}^{p-1} \left(\frac{w_i}{i+1} - w_{i+1} \right) x^{i+1} + \frac{w_p}{p+1} x^{p+1}$$

and

$$\mathscr{L}w \equiv \ell w + \frac{1}{2}. \tag{5.3-15}$$

Corresponding to the ansatz $v_i = (a_i + b_i x + c_i x^2)x^{m_i} =: m(v_i)x^{m_i}$, we set $p = m_i + 2$ and $q = m_i$ in (5.3-14). Then

$$\ell w = -\frac{1}{2}\sum_{i=m_i}^{m_i+2}\frac{w_i}{i+1} - w_{m_i} + \sum_{i=m_i}^{m_i+1}\left(\frac{w_i}{i+1} - w_{i+1}\right)x^{i+1} + \frac{w_{m_i+2}}{m_i+3}x^{m_i+3}.$$

$$\tag{5.3-16}$$

Finally setting $m(w) = m(v)$, we get

$$\ell v_i = \alpha_i(m_i) + p_i(x)x^{m_i} + \frac{c_i}{m_i+3}x^{m_i+3},$$

$$\alpha_i(m_i) := -\frac{1}{2}\left(\frac{a_i}{m_i+1} + \frac{b_i}{m_i+2} + \frac{c_i}{m_i+3}\right), \tag{5.3-17}$$

$$p_i(x) := -a_i + \left(\frac{a_i}{m_i+1} - b_i\right)x + \left(\frac{b_i}{m_i+2} - c_i\right)x^2.$$

Using (5.3-15), we summarize the equations that implement the IRC steps in the following list:

$$S_2^p(\ell v_0 + \mathscr{L}0) \equiv 0 \quad (\text{i.e., } S_2\mathscr{L}v_0 = 0), \tag{5.3-18}$$

$$S_2^p(\ell v_1 + \mathscr{L}v_0) \equiv 0 \text{ or } S_2^p(\ell v_1 + \mathscr{L}0 + \ell v_0) \equiv 0, \tag{5.3-19}$$

$$S_2^p(\ell v_2 + \mathscr{L}(v_0 + v_1)) \equiv 0 \text{ or } S_2^p(\ell v_2 + \mathscr{L}0 + \ell v_0 + \ell v_1) \equiv 0, \tag{5.3-20}$$

$$S_2^p(\ell v_3 + \mathscr{L}0 + \ell v_0 + \ell v_1 + \ell v_2) \equiv 0, \tag{5.3-21}$$

$$S_2^p(\ell v_4 + \mathscr{L}0 + \ell v_0 + \ell v_1 + \ell v_2 + \ell v_3) \equiv 0, \tag{5.3-22}$$

$$\cdots$$

$$S_2^p(\ell v_i + \mathscr{L}0 + \ell(v_0 + v_1 + \ldots + v_{i-1})) \equiv 0. \tag{5.3-23}$$

The steps corresponding to (5.3-18) and (5.3-19) (i.e., the forward corrections) have just been completed with the result being the approximation displayed in (5.3-13).

Now we proceed to solve (5.3-20) for v_2, the first of the backward corrections. By recursive use of (5.3-16), (5.3-20) may be cast into the following form:

$$S_2^p \left(\alpha_2(m_2) + p_2 x^{m_2} + \frac{c_2}{m_2+3} x^{m_2+3} + \frac{1}{2} + \alpha_0(m_0) + p_0(x)x^{m_0} \right.$$

$$\left. + \frac{c_0}{m_0+3} x^{m_0+3} + \alpha_1(m_1) + p_1(x)x^{m_1} + \frac{c_1}{m_1+3} x^{m_1+3} \right) \equiv 0.$$

Rearranging this expression we have

$$S_2^p \left(\frac{1}{2} + \alpha_0(m_0) + \alpha_1(m_1) + \alpha_2(m_2) \right.$$

$$+ \ p_0(x)x^{m_0} + p_1(x)x^{m_1} + p_2(x)x^{m_2} \qquad\qquad (5.3\text{-}24)$$

$$\left. + \ \frac{c_0}{m_0+3} x^{m_0+3} + \frac{c_1}{m_1+3} x^{m_1+3} + \frac{c_2}{m_2+3} x^{m_2+3} \right) \equiv 0.$$

So we see from the solution of (5.3-6) (which corresponds to (5.3-18)) that we have $\frac{1}{2} + \alpha_0(m_0) + p_0(x) = 0$ and $m_0 = 0$, $c_0 = \frac{3}{22}$.

Using this, (5.3-24) becomes

$$S_2^p \left(\alpha_1(m_1) + \alpha_2(m_2) + p_1(x)x^{m_1} + p_2(x)x^{m_2} \right.$$

$$\qquad\qquad (5.3\text{-}25)$$

$$\left. + \frac{c_0}{3} x^3 + \frac{c_1}{m_1+3} x^{m_1+3} + \frac{c_2}{m_2+3} x^{m_2+3} \right) \equiv 0.$$

Now for the solution of (5.3-19), we know from (5.3-9) and (5.3-10) that $m_1 = 3$, and we know the values of a_1, b_1, and c_1 as well. These give us that

$$p_1(x) + \frac{c_0}{3} \equiv 0.$$

Employing this, (5.3-25) becomes

$$S_2^p\left(\alpha_1(m_1) + \alpha_2(m_2) + p_2(x)x^{m_2} + \frac{c_1}{m_1+3}x^6 + \frac{c_2}{m_2+3}x^{m_2+3}\right) \equiv 0.$$

$$(5.3-26)$$

Setting $p = m_2 = 2$ (see Fig. 5.3-1), (5.3-26) becomes

$$S_2^p\left(p_2(x)x^2 + \frac{c_1}{6}x^6 + \frac{c_2}{5}x^5\right)$$

$$\equiv x^2 S_2\left(p_2(x) + \frac{c_2}{5}x^3 + \frac{c_1}{6}x^4\right) \equiv 0.$$

Then

$$x^2 S_2(p_2(x)) \equiv x^2 p_2(x) \equiv 0. \qquad (5.3-27)$$

Then $p_2(x) \equiv 0$, i.e., $a_2 = b_2 = c_2 = 0$ and $\alpha_2(m_2) = 0$ as well. This means that in our example (5.3-2) in this state of the IRC process there is no direct flow of information from right to the left, but rather an implicit such flow through the residue as is seen below in (5.3-32). Next for solving (5.3-21) we employ a process similar to that of (5.3-23)-(5.3-26). The current counterpart of (5.3-26) is

$$S_2^p\left(\alpha_1(m_1) + \alpha_2(m_2) + \alpha_3(m_3) + p_2(x)x^{m_2} + p_3(x)x^{m_3}\right.$$

$$(5.3-28)$$

$$\left. + \frac{c_1}{m_1+3}x^6 + \frac{c_2}{m_2+3}x^{m_2+3} + \frac{c_3}{m_3+3}x^{m_3+3}\right) \equiv 0.$$

Inserting the values $\alpha_2(m_2) = 0$, $m_2 = 2$, $c_2 = 0$, and $p_2(x) = 0$ already determined, (5.3-28) becomes

$$S_2^p\left(\alpha_1(m_1) + \alpha_3(m_3) + p_3(x)x^{m_3}\right.$$

$$(5.3-29)$$

$$\left. + \frac{c_1}{m_1+3}x^6 + \frac{c_3}{m_3+3}x^{m_3+3}\right) \equiv 0.$$

Using Fig. 5.3-2, we select $p = m_3 = 1$ for this bwd RC step. Then

$$S_2^1\left(p_3(x)x + \frac{c_3}{4}x^3 + \frac{c_1}{6}x^5\right)$$

$$\equiv xS_2\left(p_3(x) + \frac{c_3}{4}x^4 + \frac{c_1}{6}x^6\right) \tag{5.3-30}$$

$$\equiv xp_3(x) \equiv 0.$$

The last equation yields $p_3 \equiv 0$, i.e., $a_3 = b_3 = c_3 = 0$ and $\alpha_3(m_3) = 0$, that is, a null backward correction once again.

Turning now to solve (5.3-22), we proceed as for (5.3-23)-(5.3-26) and for (5.3-28)-(5.3-29). Using $\alpha_3(m_3) = c_3 = 0$, we find the following counterpart of (5.3-29):

$$S_2^p\left(\alpha_1(m_1) + \alpha_4(m_4) + p_4(x)x^{m_4}\right.$$

$$\left. + \frac{c_1}{m_1+3}x^6 + \frac{c_4}{m_4+3}x^{m_4+3}\right) \equiv 0. \tag{5.3-31}$$

Using Fig. 5.3-2 once again, we select $p = m_4 = 0$. Thus we arrive at the equation

$$\alpha_1(m_1) + \alpha_4(m_4) + p_4(x) + S_2\left(\frac{c_1}{6}x^6 + \frac{c_4}{3}x^3\right) \equiv 0.$$

Then

$$\alpha_1(m_1) + \alpha_4(m_4) + p_4(x) \equiv 0.$$

Using (5.3-16), $\alpha_1(3) = -\dfrac{67}{5280}$ (cf. (5.3-12)), so that this last equation becomes

$$\alpha_1 - \frac{1}{2}\left(\frac{a_4}{1} + \frac{b_4}{2} + \frac{c_4}{3}\right)$$

$$-a_4 + \left(\frac{a_4}{1} - b_4\right)x + \left(\frac{b_4}{2} - c_4\right)x^2 \equiv 0.$$

From this in turn we deduce the following system:

$$-\frac{3}{2}a_4 - \frac{1}{4}b_4 - \frac{1}{6}c_4 = -\alpha_1 = \frac{67}{5280},$$

$$a_4 - b_4 = 0,$$

$$\frac{1}{2}b_4 - c_4 = 0.$$

Then

$$a_4 = -\frac{67}{9680}, \quad b_4 = -\frac{67}{9680}, \quad c_4 = -\frac{67}{19360}.$$

The approximation $v = v_0 + v_1 + v_2 + v_3 + v_4 = v_0 + v_1 + v_4$ now is

$$v(x) = \frac{1}{19360}\left(5145 + 5145x + 2573x^2\right.$$

$$\left. + 880x^3 + 220x^4 + 44x^5\right). \tag{5.3-32}$$

The quality of this approximation is illustrated in Table 5.3-3.

x	$v(x)$	$y(x)$	$\left|\frac{y-v}{y}\right|$
0	0.2657	0.2689	0.012
1	0.7235	0.7311	0.011

Table 5.3-3

Comparing Tables 5.3-1 and 5.3-3, we see that the bwd RC processes have uniformized the quality of the approximation thus far obtained (in the sense of the relative error measures in those tables).

The approximation $v(x)$ in (5.3-32) represents one cycle of the IRC with carry process, the cycle consisting of two fwd steps followed by three bwd steps. This five-step IRC process may be iterated subject only to the requirement that it produces meaningful results; for instance, that it is a convergent or asymptotic process. If the solution y of the problem $\mathscr{L}y = 0$ being approximated is analytic, then a sufficient condition for the convergence of this IRC process is that the mantissas $m(v_i)$ be sufficiently long, i.e., that the $m(v_i)$ consist of sufficiently many monomials. Correspondingly, the rate of convergence increases with the mantissa length. (Standard floating-point iterations exhibit corresponding phenomena.) For the model problem, the conditions for iterating the IRC cycles are fulfilled. Corresponding to $m(v_i) = 3$, the limiting approximation is the following polynomial of degree 5:

$$v_\infty(x) = 0.268958 + 0.268958x + 0.134479x^2$$
$$+ 0.0448263x^3 + 0.0112066x^4 + 0.00224131x^5. \qquad (5.3\text{-}33)$$

The quality of the approximation of y by $v_\infty(x)$ is illustrated in Table 5.3-4.

x	$v_\infty(x)$	$y(x)$	$\left\lvert \dfrac{y - v_\infty}{y} \right\rvert$
0	0.268958	0.268941	0.000061
1	0.730669	0.731059	0.00054

Table 5.3-4

Although this table displays a good approximation, the rate of convergence of the IRC process that the table represents is poor. Increasing the number of monomials in $m(v_i)$ from three would improve this rate.

As we have seen, the number of degrees of freedom N' in the approximation v differs from the number of degrees of freedom employed by the corrections v_i, say n'. We stress that although the IRC process employs corrections v_i of a fixed (possibly a small) number of degrees of freedom n', i.e., mantissa length, the process is capable of producing approximations of arbitrarily high accuracy. To achieve this, N' but not n' must be increased and the residue $\mathscr{L}(v_0 + \ldots + v_{i-1})$ must be computed with increasing (even maximal) accuracy. Mitigating this requirement is the fact that computation of the residue represents a small part of the correction process.

validation

In order to produce an inclusion of the approximation thus far obtained, it is only necessary to perform the last step of residual correction as an inclusion step using IS_2 in place of S_2 (cf. (5.2-23)). Of course the iteration step must be rigorously performed with no terms neglected. In particular the α terms must not be neglected; as a matter of fact they are not, coincidentally, in the last step above.

5.4 A FORMALISM FOR IRC IN FUNCTION SPACE

The examples of IRC of Section 5.2 and 5.3 are effective function-space counterparts of the residual correction process. The ad hoc aspects corresponding to arithmetic operations such as rounding, carry, and the formal deletion of terms require a formalization that justifies these processes. We now supply this formalism.

To do this, we formulate a more general version of these IRC processes. Referring to (5.3-18)-(5.3-23), we consider the linear problem

$$\mathscr{L}y := \ell y + \mathscr{L}0 = 0 \qquad (5.4\text{-}1)$$

set in a separable Hilbert space $\mathscr{M}$ with basis $\Phi = (\phi_0, \phi_1, ...)$. An approximate solution of (5.4-1) is sought in a subspace $sp\Phi_M$ of $\mathscr{M}$. (M may be infinite.) The IRC process produces the approximation by solving a sequence of problems, each of dimension less than M. In Section 5.2 these subspaces corresponded to the following sequences of basis elements:

$$\begin{aligned}
\Phi^{0,N} &:= (\phi_0, \phi_1, ..., \phi_N) \\
\Phi^{N,2N} &:= (\phi_{N+1}, ..., \phi_{2N})
\end{aligned} \qquad (5.4\text{-}2)$$

$$. . .$$

along with the associated relative roundings

$$S_N^0, \ S_N^{N+1}, \ .... \qquad (5.4\text{-}3)$$

(In fact these subspaces may overlap, as in Section 5.3).

The choice of subspaces can be characterized by a sequence Γ of sequences γ_i, $i = 0, 1, \dots$:

$$\Gamma := (\gamma_0, \gamma_1, \gamma_2, \dots), \tag{5.4-4}$$

where

$$\gamma_0 = (0, 1, \dots, N),$$
$$\gamma_1 = (N + 1, \dots, 2N),$$
$$\dots \dots$$

In the more general case Γ is arbitrary except that

$$\bigcup_{i \geq 0} \gamma_i = (1, 2, \dots, M),$$

where typically each sequence γ_i has fewer than M elements.

Then let

$$\gamma_1 = (k_0, k_1, \dots, k_{N_k}), \quad k = 0, 1, \dots, \tag{5.4-5}$$

where

$$0 \leq k_i \leq M, \quad k = 0, \dots, N_k, \quad k = 0, 1, \dots .$$

The sequence Γ defines a sequence of subspaces with basis elements generalizing those chosen in (5.4-2), viz.,

$$\Phi_{\gamma_k} = (\phi_{k_0}, \phi_{k_1}, \dots, \phi_{k_{N_k}}), \quad k = 0, 1, \dots .$$

The roundings (corresponding to (5.4-3)) onto the subspaces $sp\Phi_{\gamma_k}$, are denoted by

$$S_{N_k}^{\gamma_k} : \mathcal{M} \to S_{N_k}^{\gamma_k}(\mathcal{M}) = sp\Phi_{\gamma_k}, \quad k = 0, 1, \dots . \tag{5.4-6}$$

This relative rounding is defined as follows. Let

$$y = \sum_{i=0}^{M} a_i \phi_i,$$

and let γ be a sequence of the type in (5.4-5). Then

$$S_N^\gamma y = \sum_{j \in \gamma} b_j \phi_j \, . \tag{5.4-7}$$

The b_j may be defined optimally with respect to a given norm:

$$\min_{\{b_j \mid j \in \gamma\}} \| y - S_N^\gamma y \| \, . \tag{5.4-8}$$

In fact a suboptimal choice corresponding to generalized Taylor rounding is suitable for our purposes, viz.,

$$b_j = \begin{cases} a_j, & j \in \gamma, \\ 0, & \text{otherwise} \end{cases} \tag{5.4-9}$$

The generalized IRC process for approximating a solution of (5.4-1) may be written as follows:

step 0:

    solve

$$S_{N_0}^{\gamma_0} \mathscr{L} v_0 = 0 \quad \text{for} \quad v_0 \in S_{N_0}^{\gamma_0} \mathscr{M}. \tag{5.4-10}$$

step 1:

    compute the residue

$$r_1 = S_{N_1}^{\gamma_1} (\ell v_0 + \mathscr{L} 0)$$

    solve

$$S_{N_1}^{\gamma_1} (\ell v_1 - r_1) = 0 \quad \text{for} \quad v_1 \in S_{N_1}^{\gamma_1} \mathscr{M}$$

$$\cdots$$

step $n + 1$:

    compute the residue

$$r_{n+1} := S_{N_{n+1}}^{\gamma_{n+1}} (\ell v_0 + \ell v_1 + \ldots + \ell v_n + \mathscr{L} 0) \tag{5.4-11}$$

solve

$$S^{\gamma_{n+1}}_{N_{n+1}}(\ell v_{n+1} - r_{n+1}) = 0 \quad \text{for} \quad v_{n+1} \in S^{\gamma_{n+1}}_{N_{n+1}} \mathcal{M}. \tag{5.4-12}$$
. . .

Then after m such steps the IRC process has produced the following approximation for the solution of (5.4-1):

$$v_0 + v_1 + \ldots + v_m. \tag{5.4-13}$$

The choice of the sequence Γ in (5.4-4) will control the quality of this algorithm. Examples are given in Section 5.4.2.

Remark: Except for (5.4-11) the considerations in this discussion are valid for nonlinear $\mathcal{L}$. In the nonlinear case (5.4-11) requires more substantial treatment, as is customary in nonlinear problems.

We now proceed to study the linear case of this formalism by means of its isomorphic representation in $iS_N \mathcal{M} = \mathbb{R}^M$. This will establish a connection of IRC in function spaces with the well-known process of block relaxation with steering. In Section 5.4.1 we consider a formal additive correction process in a function-space, and we then develop the isomorphic counterparts in terms of the function space basis. The counterpart is an infinite linear system, and we characterize the additive correction process as a blockwise relaxation process for that linear system. Finally, we develop a convergence criterion for this block-relaxation process. In Section 5.4.2 we show that the examples of IRC of Sections 5.2 and 5.3 treated earlier are special instances of this block-relaxation process.

5.4.1 The Block–Relaxation Formalism

In this section we develop a block-relaxation formalism for an additive correction solution process for a linear equation in a function space. A convergence criterion is also furnished.

Equation to be Solved and its Isomorphic Correspondent

Let us solve the equation (5.4-1) defined in the separable Hilbert space $\mathcal{M}$.

We seek the solution y in the form of a correction v to an approximation w, i.e.,

$$y = w + v. \qquad (5.4.1\text{-}1)$$

In the case that ℓ is linear, the equation for v is

$$0 = \ell v + \ell w + \mathscr{L}0. \qquad (5.4.1\text{-}2)$$

Let us introduce isomorphic correspondents of the functions and operators appearing in (5.4.1-2) and in terms of the basis $\Phi := (\phi_0, \phi_1, \ldots)^T$ in $\mathcal{M}$. In particular let

$$
\begin{aligned}
v &:= \Phi * v, \quad v = (v_0, v_1, \ldots)^T, \\
w &:= \Phi * w, \quad w = (w_0, w_1, \ldots)^T, \\
\mathscr{L}0 &:= \Phi * f.
\end{aligned}
\qquad (5.4.1\text{-}3)
$$

(In the case that $\mathscr{L}0$ is a constant, $f = (\mathscr{L}0, 0, 0, \ldots)^T$.) Further, let

$$\ell w = \ell \Phi * w := \Phi * \mathscr{A}(\ell)w, \qquad (5.4.1\text{-}4)$$

so that

$$\mathscr{A}(\ell) = (a_{ij})$$

is an infinite matrix that is the isomorphic correspondent to ℓ (cf. (4.1-1)).

Then the isomorphic correspondent of the equation (5.4.1-2) is

$$0 = \mathscr{A}(\ell)v + \mathscr{A}(\ell)w + f, \qquad (5.4.1\text{-}5)$$

which is a linear system (typically infinite) in the coefficient space for the unknown vector a.

Approximating Solution Procedure

We now apply the approximating solution procedure (5.4-1)-(5.4-10). For clarity and when no misunderstanding will occur, we shall omit displaying the last member of various sequences and other constructs. Consider then the sequence

$$\gamma_k := (k_0, k_1, ...) \tag{5.4.1-6}$$

introduced in (5.4-5). The norm of a sequence will denote the number of its elements so that $\| \gamma_k \|$, for instance, is the number of elements in the sequence γ_k.

Let $G_k: \mathcal{M} \to \mathcal{M}$ be the orthogonal projection onto the coefficient space corresponding to the elements $\Phi_{\gamma_k} := (\phi_{k_0}, \phi_{k_1}, ...)$. Thus for example

$$G_k v = \Phi * (0, ..., 0, v_{k_0}, 0, ..., 0, v_{k_1}, 0, ...)^T. \tag{5.4.1-7}$$

G_k is an isomorphic counterpart of the relative rounding $S_{N_k}^{\gamma_k}$ of (5.4-6) in the so-called Taylor case of (5.4-9). G_k is represented by a diagonal matrix (g_{ij}^k). That is

$$G_k v := \Phi * (g_{ij}^k) * v, \tag{5.4.1-8}$$

where, in particular,

$$g_{ii}^k := \begin{cases} 1, & i \in \gamma_k, \\ 0, & \text{otherwise}, \quad k = 0, 1, 2, ... \end{cases} \tag{5.4.1-9}$$

We shall use the same symbol G_k to denote the projection operator and its matricial correspondent.

The approximating solution procedure generates a sequence of approximations w^k, and it corresponds to a sequence Γ as follows: Let

$$w^k := G_k v^k + w^{k-1}, \quad k = 0, 1, ..., \tag{5.4.1-10}$$

with $w^{-1} \equiv 0$, and where the correction $G_k v^k$ to w^{k-1} is determined as the solution of

$$G_k(\mathcal{A}(\ell)G_k v^k + \mathcal{A}(\ell)w^{k-1} + f) = 0. \qquad (5.4.1\text{-}11)$$

Blockwise Form and Relaxation

This solution procedure (5.4.1-10) and (5.4.1-11) is precisely block relaxation with steering. The kth block is defined in turn by the kth sequence $\gamma_k = (k_0, k_1, \ldots)$. To see this and to facilitate the development of a convergence criterion for this solution procedure, we first devise a more compact representation of the process.

Setting

$$\begin{aligned}
u^k &:= G^k v^k, \\
\mathcal{B}_k &:= G_k \mathcal{A}(\ell)G_k, \\
Q_k &:= G_k \mathcal{A}(\ell), \\
g_k &:= G_k f,
\end{aligned} \qquad (5.4.1\text{-}12)$$

(5.4.1-10) and (5.4.1-11) may be written

$$\begin{aligned}
&\text{(a)} \quad w^k = u^k + w^{k-1}, \\
&\text{(b)} \quad \mathcal{B}_k u^k + Q_k w^{k-1} + g_k = 0, \quad k = 0, 1, \ldots,
\end{aligned} \qquad (5.4.1\text{-}13)$$

with $w^{-1} \equiv 0$.

(5.4.1-13) is precisely a blockwise Jacobi relaxation process corresponding to blocks of unknowns with indices in γ_k, $k = 0, 1, \ldots$. Indeed from (5.4.1-12) we see that

(i) u^k consists of those components of v^k indexed by γ_k,

(ii) $\mathcal{B}_k$ is that subblock of $\mathcal{A}(\ell)$ corresponding to those components in question,

(iii) (5.4.1-13b) solves the appropriate corresponding equations of (5.4.1-11) for the block u^k of unknowns, and

(iv) from (5.4.1-13a) the value of the current approximation w^{k-1} is augmented by this newly solved for block u^k of unknowns to yield the correction w^k.

These observations will become clearer when we consider examples.

Convergence Criterion: Periodic Case

To develop a convergence criterion for the process (5.4.1-13), we proceed with a reidentification of the quantities u^k, w^k, g_k, $\mathcal{B}_k$, and Q_k.

For example, u^k is an infinite vector with only $\|\gamma_k\|$ nonzero components. Then let $\bar{u}^k$ denote that $\|\gamma_k\|$ vector whose components are the nonzero components of u^k taken in order. These components are indexed by $\gamma_k = (k_0, k_1, ...)$, as we recall. We may view $\bar{u}^k$ as a compression, so to speak, of u_p itself. We similarly define the vector $\bar{g}_k$ as the corresponding compression of g_k. Next let $\bar{w}^k$ be the vector whose components are those components of w^k with indices in $\underset{k \geq 0}{\cup} \gamma_k$. These are all of the component indices that actually occur in the solution process as governed by the sequence γ_k. These are possibly but not necessarily all indices, 0, 1, ..., since the solution process may ignore certain components, for example, by truncation of the original system (5.4.1-5).

Next we compress the matrices $\mathcal{B}_k$ and Q_k. The nonzero components of $\mathcal{B}_k$ form a block specified by γ_k, that is, by the projection G_k as in (5.4.1-12). These are the components with row and column indices in γ_k. Let $\overline{\mathcal{B}}_k$ be the $\|\gamma_k\| \times \|\gamma_k\|$ matrix corresponding to these nonzero components of $\mathcal{B}_k$. Similarly $\overline{Q}_k$ is a matrix consisting of those rows of Q_k with indices in γ_k. The columns of $\overline{Q}_k$ consist of the columns of Q_k with indices in $\underset{k \geq 0}{\cup} \gamma_k$. Thus $\overline{Q}_k$ is a matrix of order $\|\gamma_k\| \times \underset{k \geq 0}{\Sigma} \|\gamma_k\|$.

Last, we construct the set of matrices

$$Q_{k\ell} = Q_k G_\ell, \quad k, \ell = 0, 1, 2, ..., \qquad (5.4.1\text{-}14)$$

so that compression $\overline{Q}_{k\ell}$ of $Q_{k\ell}$ itself is the block matrix whose entries correspond to the entries of $Q_{k\ell}$, rowwise to the indices in γ_k and columnwise to the indices in γ_ℓ.

Now since no confusion will result, we drop the bars, and starting with the kth equation in (5.4.1-13b), we write the equation defining the corrections as follows:

$$
\begin{aligned}
\mathscr{B}_k u^k &= -g_k - Q_k w^{k-1}, \\
Q_{k+1,k} u^k + \mathscr{B}_{k+1} u^{k+1} &= -g_{k+1} - Q_{k+1} w^{k-1}, \\
Q_{k+2,k} u^k + Q_{k+2,k+1} u^{k+1} + \mathscr{B}_{k+2} u^{k+2} &= -g_{k+2} - Q_{k+2} w^{k-1}, \\
\cdot\ \cdot\ \cdot\ &\qquad\quad \cdot\ \cdot\ \cdot
\end{aligned}
$$

$$(5.4.1\text{-}15)$$

Alternatively we may write (5.4.1-15) in terms of a lower triangular block matrix as follows:

$$
\begin{pmatrix}
\mathscr{B}_k & & \\
Q_{k+1,k} & \mathscr{B}_{k+1} & \\
Q_{k+2,k} & Q_{k+2,k+1} & \mathscr{B}_{k+2} \\
& & & \cdot
\end{pmatrix}
\begin{pmatrix}
u^k \\
u^{k+1} \\
u^{k+2} \\
\cdot
\end{pmatrix}
= -
\begin{pmatrix}
g_k + Q_k w^{k-1} \\
g_{k+1} + Q_{k+1} w^{k-1} \\
g_{k+2} + Q_{k+2} w^{k-1} \\
\cdot
\end{pmatrix}.
$$

$$(5.4.1\text{-}16)$$

Now let us suppose that the sequence Γ is periodic with period p, i.e., $\gamma_{k+p} = \gamma_k$, $k = 0, 1, \ldots$. The period p defines a cycle of p corrections $u^k, u^{k+1}, \ldots, u^{k+p-1}$ collectively called U^k. In particular let

$$
U^k = (u^k, u^{k+1}, \ldots, u^{k+p-1})^T, \quad k = 0, p, 2p, \ldots . \tag{5.4.1-17a}
$$

Similarly let

$$
F^k = (g_k, g_{k+1}, \ldots, g_{k+p-1})^T, \quad k = 0, p, 2p, \ldots , \tag{5.4.1-17b}
$$

and let

$$
R^k = (Q_k, Q_{k+1}, \ldots, Q_{k+p-1})^T, \quad k = 0, p, 2p, \ldots . \tag{5.4.1-17c}
$$

In a similar manner we define the following two matrices, D^k, a diagonal matrix, and L^k, a lower triangular matrix:

$$
D^k = \operatorname{diag}(\mathscr{B}_k, \mathscr{B}_{k+1}, \ldots, \mathscr{B}_{k+p-1}) \tag{5.4.1-18a}
$$

and

$$
L^k = \begin{pmatrix}
0 & & & & \\
\mathcal{Q}_{k+1,k} & 0 & & & \\
\mathcal{Q}_{k+2,k} & \mathcal{Q}_{k+2,k+1} & & & \\
\cdot & \cdot & \cdot & & \\
\cdot & \cdot & & \cdot & \\
\cdot & \cdot & & & \cdot \\
\mathcal{Q}_{k+p-1,k} & \mathcal{Q}_{k+p-1,k+1} & & \mathcal{Q}_{k+p-1,k+p-2} & 0
\end{pmatrix}
$$

$$(5.4.1\text{-}18b)$$

In terms of these new constructs in (5.4.1-17) and (5.4.1-18), the defining system (5.4.1-16) for the correction process becomes

$$(L^k + D^k)U^k = -F^k - R^k w^{k-1}, \quad k = 0, p, 2p, \ldots . \qquad (5.4.1\text{-}19)$$

In this equation the superscript k on the quantities K, D, F, and R serve to distinguish between choices of phase within a cycle of correction. In fact we have already made this particular choice in (5.4.1-19) by starting the process at $k = 0$. With this in mind and since no confusion will result, we rewrite (5.4.1-19) by omitting those particular superscripts. Finally, the original equation (5.4.1-13b) defining the corrections is in the form we seek, namely,

$$(L + D)U^k = -F - Rw^{k-1}. \qquad (5.4.1\text{-}20)$$

This equation is solved formally for U^k by multiplication by $(L + D)^{-1}$.

To continue to develop the convergence criterion we seek, let

$$d_j = \sum_{\ell=0}^{j} \| \gamma_\ell \|, \quad j = 0, 1, \ldots, \qquad (5.4.1\text{-}21)$$

and recall that while each u^j is a vector of size $\| \gamma_j \|$, $j = 0, 1, \ldots$, the vector U^k itself is a vector of size d_{p-1}.

Suppose the sequence Γ is a disjoint partition. That is, $\gamma_k \cap \gamma_j = \varnothing$, $k \neq j$, so that the blocks of unknowns u^k, $k = 0, 1, \ldots$ do not overlap. Then

$$U^k = \sum_{j=0}^{p-1} u^{k+j}.$$ (5.4.1-22)

That is, U^k is already a sum of the p corrections in a correction cycle so that corresponding to (5.4.1-13a), we have

$$w^{k+p-1} = w^{k-1} + U^k.$$ (5.4.1-23)

If on the other hand, the blocks of unknowns u^k do overlap, then U^k is of dimension greater than the dimension of the w^k. In this case U^k must be multiplied by an appropriate zero-one matrix K in order to produce the sum of corrections in a cycle, viz.,

$$w^{k+p-1} = w^{k-1} + KU^k.$$ (5.4.1-24)

K is a matrix of order $\| \bigcup_{j=0}^{p-1} \gamma_j \| \times (d_{p-1} - 1)$. The ones in K occur in the positions with the row indices corresponding to $\gamma_k = (k_0, k_1, \ldots)$, $k = 0, 1, \ldots, p - 1$ and column indices corresponding to $0, 1, \ldots, d_{p-1} - 1$. In particular, the ones occur at the positions

$(0_0, 0), (0_1, 1), \ldots, (0_{\| \gamma_1 \| - 1}, d_0 - 1);$

$(1_0, d_0), (1_1, d_0 + 1), \ldots, (1_{\| \gamma_2 \| - 1}, d_1 - 1);$

   (5.4.1-25)

$\ldots$

$((p-1)_1, d_{p-2} + 1), \ldots, ((p-1)_0, d_{p-2}), ((p-1)_{\| \gamma_{p-1} \| - 1}, d_{p-1} - 1).$

Here subscripted integers take their values from the sequences γ_k, $k = 0, \ldots, p - 1$ in order. The case of (5.4.1-24) subsumes that of (5.4.1-23) by setting K to the identity matrix E.

Next combining (5.4.1-20) and (5.4.1-24), we have

$$w^{k+p-1} = w^{k-1} - K(L + D)^{-1}F - K(K + D)^{-1}Rw^{k-1},$$ (5.4.1-26)

or

$$w^{k+p-1} = h + Mw^{k-1}, \qquad (5.4.1\text{-}27)$$

where

$$h = -K(L + D)^{-1}F \qquad (5.4.1\text{-}28a)$$

and

$$M = E - K(L + D)^{-1}R. \qquad (5.4.1\text{-}28b)$$

Now setting

$$t_j = w^{jp}, \quad j = 0, 1, \ldots,$$

(5.4.1-27) becomes

$$t_j = h + Mt_{j-1}. \qquad (5.4.1\text{-}29)$$

The convergence criterion we seek, namely, $\sigma(M) < 1$, may be read off from (5.4.1-28). In particular from (5.4.1-28b) it is

$$\sigma(E - K(L + D)^{-1}R) < 1. \qquad (5.4.1\text{-}30)$$

5.4.2 Application of the Relaxation Formalism

We now discuss several examples that illustrate the block-relaxation framework just developed in Section 5.4.1. First, we treat the example (5.2-1) of Section 5.2, which will show the correspondence between IRC there and the constructs of the block-relaxation process. This particular example is not iterative and so does not provide an application for the convergence criterion (5.4.1-31). This criterion is illustrated by the example (5.3-1)-(5.3-3), which we then deal with.

the example of Section 5.2

From (5.2-1) we have $\ell y = \int_0^x y - y$ and $\mathscr{L}(0) = 1$. Then using (4.1.2-2)ff, we find the isomorphic correspondent of ℓ (cf. (5.4.1-4)) is the following lower triangular matrix:

$$\mathscr{A}(\ell) = \begin{pmatrix} -1 & & & & & & \\ 1 & -1 & & & & & \\ 0 & 1/2 & -1 & & & & \\ & 0 & 1/3 & -1 & & & \\ & & 0 & 1/4 & -1 & & \\ & & & 0 & 1/5 & -1 & \\ & 0 & & & & & \cdot \cdot \cdot \end{pmatrix}. \qquad (5.4.2\text{-}1)$$

We also find that the isomorphic correspondent of f is (cf. (5.4.1-5))

$$f = (-1, 0, 0, \ldots)^T.$$

The zeroth rounded equation (5.2-2) corresponds to $\mathscr{B}_0 u^0 = g_0$ (cf. the first equation in (5.4.1-15) with $k = 0$) in correspondence to $\gamma_0 = (0, 1, 2, \ldots, N)$ (cf. (5.4.1-6)). That is, (5.2-2) corresponds to solving the first N' equations of $\mathscr{A}(\mathscr{L})u = 0$ for the first N' unknowns, $v_0, v_1, \ldots, v_N$. Indeed, when N is specialized to $N = 2$ following (5.2-5), we have $\gamma_0 = (0, 1, 2)$ so that (cf. (5.4.1-8) and (5.4.1-9))

$$G_0 = \begin{pmatrix} 1 & 0 & 0 & 0 & \\ 0 & 1 & 0 & 0 & \\ 0 & 0 & 1 & 0 & \\ 0 & 0 & 0 & 0 & \\ & & & & \cdot \cdot \cdot \end{pmatrix}. \qquad (5.4.2\text{-}2)$$

Then (cf. (5.4.1-12))

$$\mathscr{B}_0 = G_0 \mathscr{A}(\ell) G_0 = \begin{pmatrix} -1 & 0 & 0 \\ 1 & -1 & 0 \\ 0 & 1/2 & -1 \end{pmatrix}. \qquad (5.4.2\text{-}3)$$

(This is actually the compressed version $\overline{\mathscr{B}}_0$ of $\mathscr{B}_0$ with the bar dropped.) Similarly from (5.4.1-12)

$$g_0 = G_0 f = (1, 0, 0)^T. \qquad (5.4.2\text{-}4)$$

That these quantities (5.4.2-2)-(5.4.2-4) correspond to the equation for $v = a + bx + cx^2$ in Section 5.2 may be read off from the expressions following (5.2-5). Thus we have that $u^0 = (a,b,c)^T = (1,1,1/2)^T$.

Then $w^0 = (u^0, 0, 0, ...)^T$.

The first correction equation in (5.2-8), with the choice $m = 3$ made there, corresponds to solving $\mathcal{A}(\mathcal{L})v = 0$ for the block of unknowns corresponding to the powers x^3, x^4, x^5. That is, the third, fourth, and fifth equations of $\mathcal{A}(\mathcal{L})v = 0$ are solved for the third, fourth and fifth unknowns. At the same time in those equations the first three unknowns (corresponding to x^0, x^1, and x^2) are given the values previously determined for them. This in turn gives rise to the term $-1/6$ in the right member of the equation following (5.2-8). Of course, as may be seen, the terms corresponding to the remaining powers $(x^6, x^7, ...)$ do not occur in these equations since the matrix $\mathcal{A}(\ell)$ (cf. (5.4.2-1)) is lower triangular. In terms of the relaxation process, this step for the determination of the coefficients of x^3, x^4, x^5 corresponds to the sequence $\gamma_1 = (3, 4, 5)$, so that G_1 is the following matrix:

$$G_1 = \begin{pmatrix} 0 & 0 & 0 & 0 & & & \\ 0 & 0 & 0 & 0 & & & \\ 0 & 0 & 0 & 0 & & & \\ 0 & 0 & 0 & 1 & & & \\ & & & & 1 & & \\ & & & & & 1 & \\ & & & & & & \ddots \end{pmatrix},$$

where except for the three indicated ones, all entries are zero. Using G_1, we compute the quantities $Q_1 = G_1 \mathcal{A}$, $\mathcal{B}_1 = G_1 \mathcal{A} G_1$, and $g_1 = G_1 f$. In particular,

$$Q_1 = \begin{pmatrix} 0 & 0 & 1/3 & -1 & 0 & 0 & \cdot \\ 0 & 0 & 0 & 1/4 & -1 & 0 & 0 & \cdot \\ 0 & 0 & 0 & 0 & 1/5 & -1 & 0 & 0 & \cdot \end{pmatrix}.$$

Then

$$Q_1 w^0 = (1/6, 0, 0)^T.$$

$$\mathcal{B}_1 = \begin{pmatrix} -1 & 0 & 0 \\ 1/4 & -1 & 0 \\ 0 & 1/5 & -1 \end{pmatrix}, \quad g_1 = (0,0,0)^T.$$

These quantities specify the equation (5.4.1-13) for the correction u^k corresponding to $k = 1$, namely,

$$\mathcal{B}_1 u^1 + Q_1 w^0 + g_1 = 0.$$

As we have seen solving this equation corresponds to solving the third, fourth and fifth equations of $\mathcal{A}(\mathcal{L})v = 0$ for the third, fourth, and fifth unknowns (as may be read off from (5.2-8)f).

We proceed now to the more general case of (5.2-10) and (5.2-11)ff. The equation (5.2-22) corresponds to solving the equation $\mathcal{A}(\mathcal{L})v = 0$ for the ith block of unknowns. The rounding $S_{N'}^{m_i}$ in (5.2-20) corresponds to culling out the ith block of equations. For all of this we see that

$$\gamma_j = (N'j, \, N'j + 1, \, ..., \, N'j + N)$$

so that Q_j is the following matrix with N' rows:

$$\begin{pmatrix} 0 & . & . & 0 & 1/(N'j) & -1 & 0 & & . & . & . \\ 0 & . & . & 0 & 0 & 1/(N'j+1) & -1 & 0 & & & \\ . & & & & & & & & & & \\ . & & & & & . & . & . & & & \\ . & & & & & & & & & . & . \\ 0 & . & . & . & 0 & & . & . & . & 0 & 1/(N'j+N) & -1 & 0 \end{pmatrix}.$$

Here the element with value $1/(N'j)$ is the $N'j^{th}$ column, that is, the column with index $N'j - 1$. The matrices Q_{jk} (cf., (5.4.1-14)) consist of the N' columns of Q_j corresponding to the indices $N'k$, $N'k + 1$, ...

$N'k + N$. For instance

$$Q_{10} = \begin{pmatrix} 0 & . & . & . & 0 & 1/(2N') \\ 0 & . & . & . & 0 & 0 \\ & & . & . & . & \\ 0 & . & . & . & 0 & 0 \end{pmatrix},$$

Q_{20} is a zero matrix, while

$$Q_{21} = \begin{pmatrix} 0 & . & . & . & 1/(2N') \\ 0 & . & . & . & 0 \\ & & . & . & . \\ 0 & . & . & . & 0 \end{pmatrix},$$

etc.

Since $\mathscr{A}(\ell)$ in this example is lower triangular, the blockwise process is a sequential and not a recurrent process. Thus questions of convergence are vacuous in this case.

the example of Section 5.3

Having just treated the example of Section 5.2 in great detail, we content ourselves here with a summary of the particular values of the constructs of the blockwise relaxation process of Section 5.4.1 for the example of Section 5.3.   That is, we suppose that readers can now supply the necessary cross references and cross correspondents for the most part. The example of Section 5.3 corresponds to a periodic relaxation process. It is recurrent, so that more of the constructs of Section 5.4.1 will come into play.

From (5.3-2) we have

$$\ell u = - \frac{1}{2} \int_0^1 v + \int_0^x v - v,$$

$$\mathscr{L}0 = - \frac{1}{2}.$$

The period $p = 4$, and

$$\gamma_0 = (0,1,2),$$
$$\gamma_1 = (3,4,5),$$
$$\gamma_2 = (2,3,4),$$
$$\gamma_3 = (1,2,3).$$

Thus $\underset{j \geq 0}{U} \gamma_j = (0,1,2,3,4,5)$. Then the isomorphic correspondent (5.4.1-5) of equation (5.3-2) is truncated. Indeed,

$$\mathscr{A}(\ell) = \begin{vmatrix} -3/2 & -1/4 & -1/6 & -1/8 & -1/10 & -1/12 \\ 1 & -1 & 0 & 0 & 0 & 0 \\ 0 & 1/2 & -1 & 0 & 0 & 0 \\ 0 & 0 & 1/3 & -1 & 0 & 0 \\ 0 & 0 & 0 & 1/4 & -1 & 0 \\ 0 & 0 & 0 & 0 & 1/5 & -1 \end{vmatrix}.$$

The four projection operators G_j, $j = 1$, 2, 3, 4, correspond to the following diagonal matrices:

$$G_0 = \text{diag} \ (1,1,1,0,0,0),$$
$$G_1 = \text{diag} \ (0,0,0,1,1,1),$$
$$G_2 = \text{diag} \ (0,0,1,1,1,0),$$
$$G_3 = \text{diag} \ (0,1,1,1,0,0).$$

Thus

$$\mathscr{B}_0 = \begin{vmatrix} -3/2 & -1/4 & -1/6 \\ 1 & -1 & 0 \\ 0 & 1/2 & -1 \end{vmatrix}, \quad \mathscr{B}_1 = \begin{vmatrix} -1 & 0 & 0 \\ 1/4 & -1 & 0 \\ 0 & 1/5 & -1 \end{vmatrix},$$

$$\mathscr{B}_2 = \begin{vmatrix} -1 & 0 & 0 \\ 1/3 & -1 & 0 \\ 0 & 1/4 & -1 \end{vmatrix}, \quad \mathscr{B}_3 = \begin{vmatrix} -1 & 0 & 0 \\ 1/2 & -1 & 0 \\ 0 & 1/3 & -1 \end{vmatrix}.$$

For the 6×12 matrix K, we have

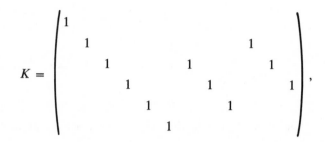

where only the nonzero entries are indicated, each column having exactly a single unit entry.

For the matrix R we have

$$
\begin{array}{|c|}
\hline
G_0 \mathscr{A} \\
\hline
G_1 \mathscr{A} \\
\hline
G_2 \mathscr{A} \\
\hline
G_3 \mathscr{A} \\
\hline
\end{array}
=
\begin{array}{|c|}
\hline
Q_0 \\
\hline
Q_1 \\
\hline
Q_2 \\
\hline
Q_3 \\
\hline
\end{array}
=
\left|
\begin{array}{cccccc}
-3/2 & -1/4 & -1/6 & -1/8 & -1/10 & -1/12 \\
1 & -1 & 0 & 0 & 0 & 0 \\
0 & 1/2 & -1 & 0 & 0 & 0 \\
\hline
0 & 0 & 1/3 & -1 & 0 & 0 \\
0 & 0 & 0 & 1/4 & -1 & 0 \\
0 & 0 & 0 & 0 & 1/5 & -1 \\
\hline
0 & 1/2 & -1 & 0 & 0 & 0 \\
0 & 0 & 1/3 & -1 & 0 & 0 \\
0 & 0 & 0 & 1/4 & -1 & 0 \\
\hline
1 & -1 & 0 & 0 & 0 & 0 \\
0 & 1/2 & -1 & 0 & 0 & 0 \\
0 & 0 & 1/3 & -1 & 0 & 0 \\
\end{array}
\right|,
$$

Further

$$
Q_{10} = \left|
\begin{array}{ccc}
0 & 0 & 1/3 \\
0 & 0 & 0 \\
0 & 0 & 0
\end{array}
\right|,
$$

$$Q_{20} = \begin{vmatrix} 0 & 1/2 & -1 \\ 0 & 0 & 1/3 \\ 0 & 0 & 0 \end{vmatrix}, \quad Q_{21} = \begin{vmatrix} 0 & 0 & 0 \\ -1 & 0 & 0 \\ 1/4 & -1 & 0 \end{vmatrix},$$

$$Q_{30} = \begin{vmatrix} 1 & -1 & 0 \\ 0 & 1/2 & -1 \\ 0 & 0 & 1/3 \end{vmatrix}, \quad Q_{31} = \begin{vmatrix} 0 & 0 & 0 \\ 0 & 0 & 0 \\ -1 & 0 & 0 \end{vmatrix}, \quad Q_{32} = \begin{vmatrix} 0 & 0 & 0 \\ -1 & 0 & 0 \\ 1/3 & 1 & 0 \end{vmatrix}.$$

Using the Q_{ij} along with the $\mathscr{B}_i$ computed above enables us to determine the matrices L and D (cf. (5.4.1-18)). Combining K, L, D, and R, we obtain M (cf. (5.4.1-28b)):

$$M = \begin{vmatrix} 0 & 5 \times 10^{-13} & -8 \times 10^{-13} \\ 0 & 0 & -8 \times 10^{-13} \\ 0 & 0 & 0 \\ 4.5 \times 10^{-13} & 7.5 \times 10^{-14} & -3.5 \times 10^{-13} \\ 5 \times 10^{-14} & 7.5 \times 10^{-14} & 8.\overline{3} \times 10^{-15} \\ 0 & 10^{-13} & -2.15 \times 10^{-13} \end{vmatrix}$$

$$\begin{vmatrix} -7.5 \times 10^{-2} & -6 \times 10^{-2} & -5 \times 10^{-2} \\ -7.5 \times 10^{-2} & -6 \times 10^{-2} & -5 \times 10^{-2} \\ -3.75 \times 10^{-2} & -3 \times 10^{-2} & -2.5 \times 10^{-2} \\ -1.25 \times 10^{-2} & 10^{-2} & -8.\overline{3} \times 10^{-2} \\ -3.125 \times 10^{-3} & 2.5 \times 10^{-3} & -2.08\overline{3} \times 10^{-3} \\ -6.25 \times 10^{-4} & 5 \times 10^{-4} & -4.1\overline{6} \times 10^{-4} \end{vmatrix}.$$

This enables us to appraise the convergence criterion (5.4.1-30). Indeed, we have for the spectral radius

$$\sigma(M) = 0.015416.$$

For purposes of comparison, we have considered the example of Section 5.3 with a number of different sequences Γ. In each case the system (5.4.1-5) is truncated to a 6×6 system; however, the word length, that is, the number of degrees of freedom in each block, N', varies. These cases are given columnwise in Table 5.4.2-1, where case 1 corresponds to the example just treated.

	1	2	3	4	5	6	7
p	4	2	4	2	3	3	6
N	2	2	2	3	3	1	0
Γ	(0,1,2)	(0,1,2)	(0,1,2)	(0,1,2,3)	(0,1,2,3)	(0,1)	(0)
	(3,4,5)	(3,4,5)	(3,4,5)	(2,3,4,5)	(4,5,0,1)	(2,3)	(1)
	(2,3,4)		(0,1,2)		(2,3,4,5)	(4,5)	(2)
	(1,2,3)		(3,4,5)				(3)
							(4)
							(5)
σ	0.01542	0.01542	0.0002377	0.002881	9.4×10^{-5}	0.0329	0.128
W	108	54	108	128	192	24	6
$\sigma^{1/W}$	0.962	0.926	0.926	0.955	0.953	0.867	0.710

Table 5.4.2-1

A rough estimate of the work W needed for one cycle of each case is simply the number of multiplications needed to solve the blocks of linear equations (cf. (5.4.1-15)) arising in such a cycle. We take this to be the sum of the cubes of the orders of the matrices $\mathscr{B}_0$, $\mathscr{B}_1$, ..., $\mathscr{B}_{p-1}$, viz.,

$$W = \sum_{j=0}^{p-1} \| \gamma_j \|^3 = p \| \gamma_0 \|^3.$$

The row labeled $\sigma^{1/W}$ in Table 5.4.2-1 shows the contraction (towards a solution) per unit of work for each of the six cases.

Chapter 6

COMMENTS ON PROGRAMMING LANGUAGE

The methodology introduced here in computer arithmetic and numerical analysis, composed as it is of many higher-level concepts and constructs, requires that a complementary higher-level programming language be furnished. Such a programming language will enable the computer user to employ these many constructs in an effective and congenial manner. Examples of programming languages that have been devised with this principle of accommodation for other classes of computer-arithmetic constructs exist and have demonstrated their effectiveness. Such examples may be found in [2], [3], [4] and [11]. For our methodology, the requirements of the higher programming language may be succinctly summarized as follows: *the language should make an efficient use of functoids and their algebraic structure possible.*

We now list some key features that our prospective programming language should have:
(i) An operator concept, such as in PASCAL-SC, ALGOL 68, and ADA.

(ii) All standard operators of the structure $(\mathcal{M}; +,\cdot,/,\int)$ should be available as standard operators. Frequently used standard functions and frequently used higher-level operators corresponding to functions should also be provided.

To illustrate the degree of congeniality that these features provide, we devise a hypothetical program in our prospective language. We do this for the following boundary-value problem, and observe that any corresponding program written in, say, FORTRAN or PASCAL would be quite involved by comparison.

boundary-value problem

$$(10 + x^2)y'' + (1 - x^2)y' + xy = x^8$$

$$y(-1) + 2y(1) = 0, \qquad (6\text{-}1)$$

$$y(0) + y'\left(\frac{1}{2}\right) = 1.$$

(6-1) is transformed into the following integral equation for $z := y''$:

$$z = \left(x^8 - x\left(a + bx + \int_0^x\int_0^x zdx^2\right)\right.$$
$$\left. - (1 - x^2)\left(b + \int_0^x zdx\right)\right)/(10 + x^2). \qquad (6\text{-}2)$$

where

$$\ell(z) := \left(\frac{1}{2} + \frac{1}{2}\int_0^{-1}\int_0^x zdx^2 + \int_0^1\int_0^x zdx^2 - \frac{1}{2}\int_0^{1/2} zdx,\right.$$
$$\left. \frac{1}{2} - \frac{1}{2}\int_0^{-1}\int_0^x zdx^2 - \int_0^1\int_0^x zdx^2 - \frac{1}{2}\int_0^{1/2} zdx\right). \qquad (6\text{-}3)$$

A number of data types will be employed.

The types *real* and *interval* are customary types, as in PASCAL-SC [3].

The new types *functoid* and *interval functoid* may be chosen as follows.

functoid is the set of all polynomials of degree N or less, the degree being adaptive.

interval functoid is the set of all interval polynomials of degree M or less, the degree being adaptive.

Because of the complexity of function space arithmetic, we require that the rounding to be used be explicitly specified. The term *rounding S* or *rounding IS* means that in all ensuing program statements, the rounding operator S resp. IS is to be applied. The explicit rounding type is characterized by the specification *Taylor(N)* resp. *interval Taylor(N)*. N specifies the degree of the rounded polynomial of a canonical memory, i.e., mantissa length.

This sample program reveals several points about the language. The programming language must recognize constants, functions, and operators of several types, such as real, polynomials, intervals, interval polynomial. Furthermore, the program has to recognize dummy variables, e.g., x for integration. These packages must fulfill certain conditions of precision, accuracy, and reliability. (See, e.g., [2]).

A listing of a hypothetical program for (6-2) and (6-3) is given. It consists of two parts: approximation and validation.

This program having run we are assured of the following three assertions concerning the solution of (6-1):

$$y'' \in Y; \quad y' \in U; \quad y \in V.$$

Sample program: approximation

var *a,b*: *real*;   *u,v*: *functoid*;

<u>mapping</u> $f(a,b,z) = \left(x^8 - x(a + bx + \int_0^x \int_0^x z\,dx) - (1 - x^2)(b + \int_0^x 2\,dx) \right) / (10 + x^2)$: *functoid*;

<u>boundarycondition</u> $\ell(z) = \left(\frac{1}{2} + \frac{1}{2}\int_0^{-1} \int_0^x z\,dx^2 + \int_0^1 \int_0^x z\,dx^2 - \frac{1}{2}\int_0^{1/2} z\,dx, \right.$

$$\left. \frac{1}{2} - \frac{1}{2}\int_0^1 \int_0^x z\,dx^2 - \int_0^1 \int_0^x z\,dx^2 - \frac{1}{2}\int_0^{1/2} z\,dx \right) : vector;$$

<u>rounding</u> *S*: *Taylor*(*N*);   "approximation step"

u := *u*;

<u>repeat</u>   *v* := *u*;

      (*a,b*) := $\ell(v)$;

      u := *f*(*a,b,v*);

<u>until</u>   norm (*u* − *v*) ≤ norm (*u*) • 10^{-8};

print(*u*);

Sample program: validation

var A,B: interval; Z,Y: interval functoid;

<u>mapping</u> $F(A,B,Z) = \left(x^8 - x(A + Bx + \int_0^x \int_0^x Zdx^2) - (1 - x^2)(B + \int_0^x Zdx) \right) / (10 + x^2)$: interval functoid;

<u>boundary</u>   <u>condition</u> $L(Z) := \left(\frac{1}{2} + \frac{1}{2} \int_0^1 \int_0^x Zdx^2 + \int_0^x \int_0^x Zdx^2) - \frac{1}{2} \int_0^{1/2} Zdx,\right.$

$\left. \frac{1}{2} - \frac{1}{2} \int_0^{-1} \int_0^x Zdx^2 - \int_0^1 \int_0^x Zdx^2 - \frac{1}{2} \int_0^{1/2} Zdx \right)$: interval vector;

<u>rounding</u>   IS: directed Taylor(N);   "validating step"
$Y := U$;

<u>repeat</u>
$Z := Y$;
$(A,B) := L(Z)$;
$Y := F(A,B,Z)$;

<u>until</u>
$\overset{\circ}{Y} \subset Z$;

$U := B + \int_0^x Y(x)dx$

$V := A + Bx + \int_0^x \int_0^x Y(x)dx^2$

print(Y,U,V);

Chapter 7

APPLICATION AND
ILLUSTRATIVE COMPUTATION

The implementation of self-validating numerics on a computer followed by illustrative numerical experiments is the subject of this chapter.

In Section 7.1 we give a concise review of the self-validating technique. The review is relatively self-contained. It emphasizes the issues of implementation dealing with the passage from theoretical computation to computation on a screen (i.e., to computation on a computer). A commentary on the practical implementation of the ultra-arithmetic is also given.

In Section 7.2 we give a set of illustrative example computations. The linear case is treated first, then the nonlinear case; finally computations illustrating IRC are given.

7.1 REVIEW OF THE COMPUTATIONAL PROCESS

As a guide to the practical implementation of self-validating numerics, we review the mapping techniques that we have developed in a concise form. Although we are concerned with function-space problems, the ideas are most easily followed by changing the setting from $\mathcal{M}$ to $\mathbb{R}$. Thus we begin in Section 7.1.1 with a review of our processes in the setting $\mathbb{R}$. Then in Section 7.1.2 we review the ideas once again, this time in the setting $\mathbb{R}^N$. The treatment in $\mathbb{R}^N$ serves more than an expositional objective, since computation corresponding to the setting $\mathcal{M}$ itself is frequently conducted in the isomorphic image $\mathbb{R}^N$ of the screen $S_N(\mathcal{M})$. Finally, the review of the computational process for the function space case $\mathcal{M}$ is given in Section 7.1.3. In Section 7.1.4 a commentary on practical issues of implementation of ultra-arithmetic is given.

7.1.1 Validation in $\mathbb{R}$

Let $f(x)$ be a continuous function. We seek to solve the equation

$$x = f(x) \tag{7.1.1-1}$$

set in a domain $D \subset \mathbb{R}$, and we do so by performing the iteration

$$y_{i+1} = f(y_i), \quad i \geq 0, \tag{7.1.1-2}$$

$y_0 \in \mathbb{R}$ given. Our point of view is an a posteriori one. That is, the iteration is conducted until an indication is given that a fixed point x has been approximated in a sense to be made precise, or that because of a resource limitation (i.e., an arbitrarily specified bound on i is reached), the iteration is aborted. This being the case, it is advisable, although not mandatory, that some likelihood of success for the iteration (7.1.1-2) be known. For example, it may be known or known to be likely that f is contracting on D.

Now in fact the iteration in (7.1.1-2) is to be performed on a computer, so that (7.1.1-2) is replaced by

$$v_{i+1} := \tilde{f}(v_i), \quad i \geq 0, \qquad (7.1.1-3)$$

$v_0 \in S(\mathbb{R})$ given. $S(\mathbb{R})$ is the floating-point system of a computer, where $S: \mathbb{R} \to S(\mathbb{R})$ is some floating-point analogue of the projection operator S_N (cf. (3.1-0)). That is, S is a monotone antisymmetric rounding (cf. [1], [11]). $\tilde{f}$ is a computer defined version (i.e., an approximation) of f. That is, $\tilde{f}: S(\mathbb{R}) \to S(\mathbb{R})$.

$\tilde{f}$ may be defined by semimorphism following (3.1-6)). That is,

$$\bigwedge_{x \in S(\mathbb{R})} \tilde{f}(x) := \square f(x) := S(f(x)). \qquad (7.1.1-4)$$

Taking $\tilde{f}(x)$ in this way is optimal since the monotonicity of S implies that there is no floating-point number between $\tilde{f}(x)$ and $f(x)$. Alternatively consider the case that f is a rational function of x. We indicate this by writing

$$f(x) := f(x; +, -, \bullet, /). \qquad (7.1.1-5)$$

In this case $f(x)$ is defined as follows:

$$\bigwedge_{x \in S(\mathbb{R})} \tilde{f}(x) := f(x; \boxplus, \boxminus, \boxdot, \boxslash), \qquad (7.1.1-6)$$

indicating that $\tilde{f}(x)$ is the same rational function of x that $f(x)$ is except that the arithmetic operations are replaced by their computer-defined analogues, the latter being defined by semimorphism (cf. (3.1-6)).
This choice for $\tilde{f}(x)$ is usually far easier to obtain than (7.1.1-4) and, not surprisingly, is usually suboptimal. For nonrational functions such as $f = \sin x$, the appropriate approximation $\tilde{f}$ must be obtained by special independent means.

With the meaning of the computer iteration (7.1.1-3) specified, we execute it, say k times, to produce an approximation of an uncertain quality, say v_k, of the fixed point x. Our objective is to exploit v_k to produce by computation, and when possible, more information about the fixed point x of (7.1.1-1) itself namely,

  (i) a statement that x exists,

  (ii) a statement that x is unique, and

  (iii) error bounds for x of high quality.

To achieve this type of additional information about x from the given approximation $v := v_k$, we make use of self-validation methods. The procedure by means of which this is accomplished is called an E-process (cf. Section 2.1). The statements (ii) and (iii) are shown by a technique of local linearization. We shall then in fact use this local linearization to furnish a fourth property:

  (iv) the effective treatment of iteration operators of so called New-ton type (which frequently occur in numerical methods).

(i) VALIDATION OF EXISTENCE

Suppose, for example, we seek to establish the existence of x and to obtain an error bound (not necessarily of high quality) as well. We may employ the Schauder-Tychonoff theorem (cf. Section 2.1) to do this. (Since now the setting is $\mathbb{R}$, the Brouwer fixed point theorem suffices.) We replace (7.1.1-2) by the following iteration defined on the power set $P\mathbb{R}$ of $\mathbb{R}$, indeed defined usually on an interval space $I\!I\mathbb{R} \subseteq P\mathbb{R}$:

$$Y_{i+1} = F(Y_i), \quad i \geq 0, \tag{7.1.1-7}$$

with $Y_0 = v$.

To perform this iteration, we define the interval valued interval function extension F of f as

$$\bigwedge_{x \in X \in I\!I \mathbb{R}} f(x) \in F(X). \tag{7.1.1-8}$$

$F(X)$ may well be defined by taking it to be minimal with respect to the property indicated in (7.1.1-8). That is, by setting $F(X)$ to be the intersection of all intervals that contain $\{f(x) \mid x \in X \in I\!I\mathbb{R}\}$. This is clearly an optimal choice for $F(x)$.

In fact the iteration (7.1.1-7) is not defined on the computer, and so, we must replace it by the following one, which is so defined:

$$Z_{i+1} = \tilde{F}(Z_i), \quad i \geq 0, \qquad (7.1.1-9)$$

with $Z_0 = v$.

$\tilde{F}$ is a *computer-defined version* of F. The optimal definition of $\tilde{F}$ is obtained by employing semimorphism once again: This time we employ a mapping $IS: P\mathbb{R} \to I\!S(P\mathbb{R})$ where IS is a floating-point analogue to IS_N. More specifically, analogous to (3.2-1), we have

$$\bigwedge_{Z \in I\!I \mathbb{R}} \tilde{F}(Z) = \Diamond F(Z) := ISF(Z), \qquad (7.1.1-10)$$

in terms of the directed rounding $\Diamond := IS$ (cf. (3.2-5)).

The property of isotoney for $\Diamond$ implies the optimality of this choice for $\tilde{F}$, namely, that $\tilde{F}$ is the intersection of all computer-representable intervals that contain $F(Z_i)$ (cf. [11]).

As with $\tilde{f}$ suboptimal but simpler to obtain versions of the function $\tilde{F}(Z)$ are available in the case that F is a rational function, that is, in the case that

$$F(Z) = F(Z; +, -, \bullet, /). \qquad (7.1.1-11)$$

In this case we take

$$\bigwedge_{Z \in IS(\mathbb{R})} \tilde{F}(Z) := F(Z; \diamondsuit, \diamondsuit, \diamondsuit, \diamondsuit). \qquad (7.1.1-12)$$

In fact, to go even further, $\tilde{F}$ may be chosen to be any function from $IS(\mathbb{R})$, the set of real intervals representable on the computer into itself, with the following property:

$$\bigwedge_{Z \in IS(\mathbb{R})} \Diamond F(Z) \subset \tilde{F}. \qquad (7.1.1-13)$$

Such suboptimal choices $\widetilde{F}$ are frequently adequate, but the quality of the choice affects the computation, as may be expected.

We shall call a computer-defined version (7.1.1-9) of (7.1.1-7) wherein the property in (7.1.1-13) is supplied an *isotonic (computer) implementation* of the iteration (7.1.1-7).

Now the isotonic implementation (7.1.1-9) is executed until (if ever) an index $i = k$ is reached such that $Z_{k+1} \subseteq Z_k$, equivalently[†] such that

$$\widetilde{F}(Z_k) \subseteq Z_k. \qquad (7.1.1-14)$$

Combining this with (7.1.1-13) gives

$$F(Z_k) \subseteq Z_k. \qquad (7.1.1-15)$$

Combining this in turn with (7.1.1-8), we get

$$\bigwedge_{x \in Z_k} f(x) \in Z_k. \qquad (7.1.1-16)$$

Thus f maps Z_k into itself. Then as an appeal to the Brouwer fixed point theorem shows, the computation (7.1.1-9) has demonstrated the existence of a fixed point x of (7.1.1-1) and has produced the bound $x \in Z_k$ as well.

(ii) VALIDATION OF UNIQUENESS

With the computation having established the existence of the fixed point x (and a bound, possibly crude, as well) we continue by making a computation demonstrating the uniqueness of $x \in Z_{k+1}$. To do this, we shall make use of another of the fixed point theorems of Chapter 2, such as Theorem 4 in Section 2.2.1, reduced of course, to the context of the setting in $\mathbb{R}$ now under consideration. Then according to (2.2.1-1), we presume a set $\mathscr{K}(Y)$ of linear functions of $\mathbb{R} \to \mathbb{R}$ is given. The set is

---

[†] For techniques to enhance achieving this computational containment property, see [9].

parameterized by the interval $Y \in I\!I\!R$. Now we define the function $H(y;Y): \mathbb{R} \rightarrow I\!I\!R$ as follows: Let g and z in $\mathbb{R}$ be given, and let $Y \in I\!I\!R$ be given. Then

$$\bigwedge_{y \in Y} f(y) \in g + \mathcal{K}(Y)(y - z) =: H(y;Y). \qquad (7.1.1\text{-}17)$$

Remark: $H(y;Y)$ will be subsequently employed only in the case that y is set equal to Y. The resulting function $H(Y;Y)$ is for convenience written simply as $H(Y)$. Anticipating this we shall hereafter write $H(y;Y)$ simply as $H(y)$, since no confusion should result.

In the case that $f \in C^1(D)$ and $Y \subset D$, we may use the mean-value theorem to produce such a function H. Indeed, write

$$f(y) = f(v) + f'(\tau)(y - v), \qquad (7.1.1\text{-}18)$$

which holds for all $v, y \in D$ and an appropriate $\tau \in y \,\underline{u}\, v$ ($\underline{u}$ denotes convex union). Here τ depends on y and v. Using (7.1.1-18), we see that the mapping in (7.1.1-17) may be composed as follows: $g := f(v)$, $z := v$, $\mathcal{K}(Y) := f'(Y \,\underline{u}\, v)$, and so,

$$H(y) := f(v) + f'(Y \,\underline{u}\, v)(y - v). \qquad (7.1.1\text{-}19)$$

Now following Theorem 4 (or cf. [9]), we take the following explicit form for the iteration process (7.1.1-7):

$$Z_{i+1} = H(Z_i) := f(v) + f'(Z_i \,\underline{u}\, v)(Z_i - v), \quad i \geq 0, \qquad (7.1.1\text{-}20)$$

with $Z_0 := v$, the latter interpreted as an element of $I\!I\!R$. The iteration is to be repeated until (if ever) $Z_{k+1} \overset{\circ}{\subset} Z_k$. Let $Z := Z_{k+1}$.

We claim that

$$\bigwedge_{\tau \in Z_k} f(\tau) \in Z = f(v) + f'(Z_k \,\underline{u}\, v)(Z_k - v) \overset{\circ}{\subset} Z_k. \quad (7.1.1\text{-}21)$$

This being the case, f maps Z_k into its interior, so that a fixed point x exists, and by Theorem 4, this fixed point is unique. Let us verify the claim (7.1.1-21).

Since $\tau \in Z_k$ implies that $\tau \underline{\cup} v \subset Z_k \underline{\cup} v$, we may use the mean value theorem (7.1.1-18) to deduce that

$$f(\tau) \in \bigcup_{\sigma \in Z_k \underline{\cup} v} (f(v) + f'(\sigma)(\tau - v))$$

$$= f(v) + \left(\bigcup_{\sigma \in Z_k \underline{\cup} v} f'(\sigma) \right)(\tau - v). \qquad (7.1.1\text{-}22)$$

Now let T be an interval, and let $g(T)$ be the interval extension of a function $g(t)$. Then (cf. [1])

$$\bigcup_{\tau \in T} f'(\tau) \subset f'(T). \qquad (7.1.1\text{-}23)$$

Combining (7.1.1-22) and (7.1.1-23), we obtain

$$f(\tau) \in f(v) + f'(Z_k \underline{\cup} v)(\tau - v). \qquad (7.1.1\text{-}24)$$

Replacing $\tau - v$ in the right member here by the set $Z_k - v$ containing it, we have, a fortiori, $f(\tau) \in Z$, demonstrating our claim.

The argument for producing uniqueness by computation is not yet complete, since the iteration (7.1.1-20) establishing the uniqueness is not defined in a computer. To complete this, we must make a passage from this iteration $Z_{i+1} = H(Z_i)$ to an iteration $Z_{i-1} = \tilde{H}(Z_i)$ defined in the computer. This is done exactly as in the passage from $Y_{i+1} = F(Y_i)$ in (7.1.1-7) to $Z_{i+1} = \tilde{F}(Z_i)$ in (7.1.1-9) using one of the three modes of extension discussed following (7.1.1-9). Then the iteration $Z_{i+1} = \tilde{H}(Z_i)$ is conducted in the computer until (if ever) $Z_{k+1} \overset{\circ}{\subset} Z_k$. Then the argument used in conjunction with (7.1.1-14) and (7.1.1-15) shows that $H(Z_k) \overset{\circ}{\subset} Z_k$. This last detail completes the demonstration of obtaining uniqueness by a computation.

(iii) BOUNDS OF HIGH QUALITY

Bounds of high quality are obtained by applying IRC to the iteration. As we have seen in Chapter 5, a linear form of the iteration process makes for a straightforward application of IRC. We apply IRC to the iteration $Z_{i+1} = H(Z_i)$ in (7.1.1-20). When the problem (7.1.1-1) is of such a nature that the correction mechanism of IRC is effective for (7.1.1-20), then the containing interval produced is corrected, i.e., refined. That is, the bound $x \in Z_k$ is made sharper.

In particular, let us introduce the correction $U_i := Z_i - v$ of v into (7.1.1-20). Then

$$U_{i+1} := U_0 + f'((U_i + v) \underline{\cup} v)U_i, \quad i \geq 0, \qquad (7.1.1-25)$$

with $U_0 := f(v) - v$, the residue of the given equation (7.1.1-1) at $x = v$. Of course, as in (i) and (ii) the iteration (7.1.1-25) must be replaced by an isotonic computer implementation of itself. Iteration is then conducted until (if ever) $U_{k+1} \overset{\circ}{\subset} U_k$. Let $U := U_{k+1}$. Then exactly as in (iii) we may show that $U + v$ exists and is unique. Note that the addition $U + v$ here of screen elements $U \in IS_N(P\mathcal{M})$ and $v \in S_N(\mathcal{M})$ is not the computer addition $\diamondsuit$, but is, as indicated, exact addition. This exact addition may be performed by employing the so-called exact inner product (cf. [1]).

The simple variation (7.1.1-25) of the fixed point iteration process, being in the IRC form, furnishes the benefits of improvement of accuracy, when, in fact, these benefits are available.

(iv) EFFECTIVE TREATMENT OF OPERATORS OF THE NEWTON TYPE

Suppose that the fixed point equation (7.1.1-1) is of the so-called Newton type:

$$f(x) = x - \mathcal{D}g(x), \qquad (7.1.1-26)$$

where $g: \mathbb{R} \to \mathbb{R}$ is differentiable (cf. (2.1-7)). Then use of the direct interval extension

$$Y_{i+1} := Y_i - \mathcal{D}g(Y_i), \quad i \geq 0, \qquad (7.1.1-27)$$

$Y_0 = 0$, is not effective for computations of the nature that are performed in (i), (ii), and (iii). Indeed effectiveness in (i), (ii), and (iii) requires that an iteration based on (7.1.1-27) produce a sequence of intervals tending to zero. The computational treatment of (7.1.1-27) requires dealing with the two occurrences of the interval Y_i in its right member separately, a worst-case treatment. Letting $d: I\!\!R \rightarrow I\!\!R^+$ be a mapping supplying the diameter of an interval, we have from (7.1.1-27) that

$$d(Y_{i+1}) = d(Y_i) + |\mathscr{D}| d(g(Y_i)).$$

Here $|\mathscr{D}|$ denotes some norm of $\mathscr{D}$. Since $|\mathscr{D}| d(g(Y)) > 0$, the iteration (7.1.1-27) cannot be expected to produce a sequence of intervals tending to zero.

A local linearization of (7.1.1-26) obtained through use of the mean-value theorem,

$$f(x) = v - \mathscr{D}g(v) + (1 - \mathscr{D}g'(\tau))(y - v) \qquad (7.1.1-28)$$

with $\tau \in y \underline{\cup} v$, is effective for dealing with the difficulty in question (cf. (4.1.1-8)). By using (7.1.1-28), we replace (7.1.1-27) by

$$Y_{i+1} := v - \mathscr{D}(g(v) + (1 - \mathscr{D}g'(Y_i \underline{\cup} v))(Y_i - v), \quad i \geq 0, \qquad (7.1.1-29)$$

with $Y_0 = v$. The right member of this iteration has two occurrences of the interval Y_i also. However, these appear in a product form, and so, we do not preclude the production by (7.1.1-29) of a sequence of intervals that tend to zero. We also note that (7.1.1-29) leads to a validation that proves that $\mathscr{D}$ and g' are non singular (cf. Remark 0 following Theorem 3 in Section 2.2).

7.1.2 Validation in $I\!\!R^N$

Consider now the case in which $x \in I\!\!R^N$ and $f: I\!\!R^N \rightarrow I\!\!R^N$. The discussion of Section 7.1.1 carries over to this case practically verbatim, with these formal changes from $I\!\!R$ to $I\!\!R^N$ being made: iterations, intoness, linearizations, residual corrections as well as appropriate isotonic computer implementations are all meaningful notions in the vectorial case. The only exception to this is the use of the mean-value theorem, which is not valid in $I\!\!R^N$ in the form given in (7.1.1-18) for the construction of the particular local linearization represented by the function $H(y)$ in (7.1.1-17). This local linearization was employed in the case of iterations on intervals used for instance in Section 7.1.1(i) and (ii).

To deal with this exception, we replace the mean value theorem by Taylor's theorem with remainder:

$$f(y) = f(v) + \int_0^1 \partial_y f(\alpha y + (1 - \alpha)v)*(y - v)d\alpha. \quad (7.1.2-1)$$

Here $\partial_y: f \to \left(\dfrac{\partial f}{\partial y_1},..., \dfrac{\partial f}{\partial y_n} \right)$ is the gradient operator, and $*$ as is usual, denotes the scalar product in $\mathbb{R}^N$.

Now let $U \subset \mathbb{R}^N$ be any set with the property $y \underline{u} v \subset U$, i.e., $\alpha y + (1-\alpha)v \in U$ for all $\alpha \in [0, 1]$. Then using (7.1.2-1), we get

$$f(y) \in f(v) + \int_0^1 \partial_y f(U)*(y - v)d\alpha. \quad (7.1.2-2)$$

Since the integral here is independent of α, we rewrite (7.1.2-2)

$$f(y) \in f(v) + \partial_y f(U)*(y - v). \quad (7.1.2-3)$$

Now $Y \in \mathbb{IR}^N$, and such that $y \in Y$. Then a particular admissible choice for U in (7.1.2-3) is $U = Y \underline{u} v$.

Now corresponding to the local linearization procedure (7.1.1-17), we may take

$$\mathcal{K}(Y) := \begin{pmatrix} \partial_y f_1(Y \underline{u} v) \\ ... \\ \partial_y f_N(Y \underline{u} v) \end{pmatrix}.$$

We note that this choice of $\mathcal{K}(y)$ is the extension $\dfrac{\partial f}{\partial y}(Y \underline{u} v)$ to $P\mathbb{R}^N$ of the Jacobian of $f(y)$.

For actual computation, $\mathcal{K}(y)$ must be replaced by an $N \times N$ matrix-valued isotonic computer implementation

$$\tilde{\mathcal{K}}(Y): \mathbb{IR}^N \to \mathbb{IR}^N \times ... \times \mathbb{IR}^N.$$

This replacement may be performed by a componentwise (i.e., suboptimal) version of the corresponding isotonic computer implementation in Section 7.1.1(i) discussed following (7.1.1-24). This componentwise approach to the computer implementation is convenient but not mandatory. A more global, even optimal implementation is to be preferred. However, as we shall see later in this chapter, the actual computations are performed componentwise (for instance, the intoness is verified componentwise). Thus while this componentwise implementation is suboptimal, it is, at least for now, the operative method.

7.1.3 Validation in $\mathcal{M}$

We turn now to the setting $\mathcal{M}$, a Banach space. We seek a fixed point $y \in \mathcal{M}$ of the mapping $f \colon \mathcal{M} \to \mathcal{M}$. As in the case of the setting $\mathbb{R}^N$ discussed in Section 7.1.2, we observe that most of the discussion of Section 7.1.1 for the setting $\mathbb{R}$ carries over to $\mathcal{M}$. Once again the local linearization (cf. (7.1.1-17)) needs special scrutiny. Here, compared to the case of $\mathbb{R}^N$, our discussion goes further. We make some specific comments about the isotonic implementation of the local linearization to be developed. Following that, the special feature of the function-space case, namely, that computation is performed in the isomorphic image $IR_N(P\mathcal{M})$ of the function-space screen $IS_N(\mathcal{M})$, is dealt with.

To begin, consider the question of the local linearization. The mapping f must be such that (7.1.1-17) holds. Namely, there exists g and z in $\mathcal{M}$ as well as a set of linear operators $\mathcal{K}(Y) \colon P\mathcal{M} \to P\mathcal{M}$ such that

$$\bigwedge_{y \in Y} f(y) \in g + \mathcal{K}(Y)(y - z) =: H(y) \qquad (7.1.3\text{-}1)$$

holds. (Recall the remark following (7.1.1-17) concerning $H(y)$.)

We consider the case where $\mathcal{M}$ is a Banach space and where $f(y) \colon \mathcal{M} \to \mathcal{M}$ has a Frechet derivative $\partial_y f(y)$. Then Taylor's theorem with remainder (7.1.2-1) is valid in $\mathcal{M}$. Now the argument concerning equations (7.1.2-1)-(7.1.2-3) may be carried over into $\mathcal{M}$, replacing $U \subset \mathbb{R}^N$ by a $U \in \mathcal{M}$ and by dropping the asterisk.

In this manner we obtain

$$f(y) \in f(v) + \mathcal{K}(Y)(y - v) =: H(y) \qquad (7.1.3\text{-}2)$$

where

$$\mathcal{K}(Y) = \partial_y f(Y \underline{\cup} v).$$

The following example, which constructs $\mathcal{K}(y)$ directly, will illuminate the meaning and use of (7.1.3-2). Let $p(x,t,\phi,\eta)$: $\mathbb{R}^4 \to \mathbb{R}$ be given and consider the specific mapping

$$f(y) := \int_0^x p(x,t,y(t),y(x))dt \qquad (7.1.3\text{-}3)$$

of $\mathcal{M} \to \mathcal{M}$. Under appropriate smoothness conditions on p, we apply the mean value theorem to it:

$$p(x,t,\phi,\eta) = p(x,t,v(x),v(t))$$

$$(7.1.3\text{-}4)$$

$$+ p_3(x,t,\sigma_1,\sigma_2)(\phi - v(t)) + p_4(x,t,\sigma_1,\sigma_2)(\eta - v(x)).$$

Here p_3 and p_4 denote the partial derivatives of p with respect to its third and fourth arguments, respectively. $(v(t),v(x)) \in \mathbb{R}^2$ and $(\sigma_1,\sigma_2) \in \mathbb{R}^2$ is an appropriate point in the line segment

$$(\phi,\eta) \underline{\cup} (v(x),v(t)).$$

(7.1.3-4) holds for each particular pair (x,t) in a domain $D \in \mathbb{R}^2$ of definition of p wherein also smoothness properties giving the mean value theorem are taken to hold as well.

Now let $Y \in \mathbb{LM}$, let $y \in Y$, and let ϕ and η be particular values of y. Then using (7.1.3-4) and the argument demonstrating (7.1.1-21), we have

$$f(y) \in \int_0^x p(x,t,v(t),v(x))dt$$

$$+ \int_0^x p_3(x,t,Y(t) \underline{\cup} v(t), Y(x) \underline{\cup} v(x))(y(t) - v(t))dt$$

$$+ \int_0^x p_4(x,t,Y(t) \underline{\cup} v(t), Y(x) \underline{\cup} v(x))(y(x) - v(x))dt$$

$$:= g + \mathcal{K}(Y)(y - v)$$

$$:= H(y), \tag{7.1.3-5}$$

where $g := f(v) = \int_0^x p(x,t,v,(t),v(x))dt$. (Recall the remark concerning $H(y)$ following (7.1.1-17).) This concludes the example.

Let us now employ (7.1.3-2) in the customary manner: we replace y in (7.1.3-2) by a set $Z \in \mathbf{P\mathcal{M}}$, and the we iterate the following corresponding recurrence:

$$Z_{k+1} = g + \mathcal{K}(Z_k)(Z_k - v), \tag{7.1.3-6}$$

producing a sequence in $\mathbf{P\mathcal{M}}$.

If for some k we achieve the relation

$$Z_{k+1} \overset{\circ}{\subset} Z_k, \tag{7.1.3-7}$$

we deduce from (7.1.3-2) that $f(y) \in Z_k$, the intoness property. Then from Theorem 4 in Section 2.2.1 we have the existence and uniqueness of the fixed point y of $f(y)$ in Z_k. From the linear form of $H(y)$ here, we conclude as in Section 7.1.1(iii) that IRC methods are available for attempting to refine the quality of the bound Z_k of the fixed point.

Let us now describe an isotonic screen implementation of the recurrence (7.1.3-6). That is, a recurrence producing a sequence in $IS_N(\mathbf{P\mathcal{M}})$, an element of the new sequence containing the corresponding element of the former.

We refer to the theory in Section 3.2. Let

$$f \in (\mathcal{M}; +, -, \bullet, /, \smallint).$$

In this case for $H(y)$ defined in (7.1.3-2), we have $H(y) = H(y; +, -, \bullet, /, \smallint)$. More explicitly, separating the dependence of H on y and on Y we have $H(y) = H(y, Y; +, -, \bullet, /, \smallint)$, so that

$$H(y) \in (\mathbf{P}\mathcal{M}; +, -, \bullet, /). \qquad (7.1.3-8)$$

In terms of the rounding $IS_N: \mathbf{P}\mathcal{M} \to IS_N(\mathbf{P}\mathcal{M})$, the isotonic extension $\tilde{H}(y): IS_N(\mathbf{P}\mathcal{M}) \to IS_N(\mathbf{P}\mathcal{M})$ may be obtained in three different ways (compare (7.1.1-10), (7.1.1-12), and (7.1.1-13)):

1. $\quad \tilde{H}(Z) = IS_N H(Z).$

This is the optimal definition analogous to (3.2-1) in Section 3.1 (compare (7.1.1-10) with the rounding $\Diamond$ there replaced by IS_N here).

2. $\quad \bigwedge_{Z \in IS_N(\mathbf{P}\mathcal{M})} \tilde{H}(Z) = H(Z; \oplus, \ominus, \odot, \Diamond, \oint).$

This is a suboptimal choice, which is available in the case that $H(y)$ has the form indicated by (7.1.3-9) (compare (7.1.1-12)). This choice uses the concepts described in Section 3.2 (cf. (3.2-5)f.).

3. $\quad \bigwedge_{Z \in IS_N(\mathbf{P}\mathcal{M})} IS_N H(Z) \subset \tilde{H}(Z).$

This is simply the defining property of an isotonic computer implementation (compare (7.1.1-13)f.).

isomorphic representations

Referring to Section 3.2, we know that all elements F in the screen $IS_N(\mathbb{L}\mathcal{M})$ have the following representation:

$$F = IS_N(F) = \Phi_N * IR_N F \qquad (7.1.3-9)$$

(cf. (3.2-3)). This is the situation in which F is characterized by finitely many degrees of freedom. Suppose in addition that H has the form required for the isotonic screen extension in 2 above, (i.e., $\tilde{H}$ is an element of the functoid $(IS_N(P\mathcal{M}); \diamondsuit, \diamondsuit, \diamondsuit, \diamondsuit, \oint)$). In this case $\tilde{H}$ is well-defined on the screen $IS_N(P\mathcal{M})$ of elements which have the form (7.1.3-9). Then the operations of which $\tilde{H}$ is composed are executable in the interval vector space $(IR_N(P\mathcal{M}); +,-,*)$ to which $IS_N(P\mathcal{M})$ is isomorphic (cf. (3.2-5)f.). In particular, the verification of the inclusion (7.1.3-8) is made as in (3.2-8). That is, (7.1.3-8) is verifiable componentwise or coefficientwise.

Thus the computational methods for $\mathcal{M}$ are reduced to computations in the structure $(IR_N(P\mathcal{M}); +,-,*)$. This is precisely a setting of the form just treated in Section 7.1.2. Thus for the last stage of implementation here, namely, the isotonic computer implementation, we refer to Section 7.1.2.

7.1.4 Implementation of the Ultra–arithmetic

In this section, we comment on techniques for the practical implementation of the ultra-arithmetic in functoids and in interval functoids. The operations of ultra-arithmetic are defined through semimorphism as described in (3.1-8) for functoids and in (3.2-5) for interval functoids. Practical implementation, however, requires that appropriate approximations to semimorphism be made, as we shall see.

$sp\ \Phi_N$ is invariant under $+$ and $-$. For clarity, suppose for the remainder of this section that Φ is a polynomial basis. Then $\cdot$ leads to an element in $sp\ \Phi_{2N}$, while $\int$ leads to an element in $sp\ \Phi_{N+1}$. The rounding S_N then leads directly to the required semimorphic result. Similar observations prevail for the case of $Isp\ \Phi_N$ and the rounding IS_N.

In our implementation to follow, we stress that the properties (3.1-8) and (3.2-5) of semimorphism are not usually exactly satisfied. They are satisfied for $+$ and $-$, but not usually for $\cdot$ and $/$. The reason is a practical one, since computing the best approximations required for

$$p \boxdot q = S_N(p \cdot q),$$

$$p \Diamond\!\!\!\!\odot\, q = IS_N(p \cdot q),$$

$$p \boxslash q = S_N(p/q),$$

$$p \Diamond q = IS_N(p/q), \qquad\qquad (7.1.4\text{-}1)$$

demands a great deal of work for every operation. Since our computations in functoids and in interval functoids are iterative and proceed by refinement, it is unnecessary to achieve sharp values demanded by (7.1.4-1) in every step. It is more sensible to gain higher accuracy through further iteration steps such as those of the IRC processes described in Chapter 5. Of course, with respect to validation the implemented operations in the interval functoid must be isotonic. That is, in place of (7.1.4-1) we must have

$$p \Diamond\!\!\!\!\odot\, q \supset IS_N(p \cdot q),$$

$$p \Diamond q \supset IS_N(p/q),$$

etc.

Let p, q, and u be polynomials in $sp\ \Phi = S_N(\mathscr{M})$.

$$p(x) = \sum_{i=0}^{N} p_i x^i, \qquad q(x) = \sum_{i=0}^{N} q_i x^i, \qquad (7.1.4\text{-}2)$$

and $u(x) := p(x) \boxdot q(x)$, $\circ \in \Omega$, where

$$u(x) = \sum_{i=0}^{N} u_i x^i = \Phi_N \cdot iu \in S_N(\mathscr{M}) \qquad (7.1.4\text{-}3)$$

To implement the computer operation $\boxdot$ we shall make use of the rounding matrices $\mathscr{A}(S_N)$ (cf. (3.1-2)) and $\overset{\wedge}{\mathscr{A}}(S_N)$ (cf. (3.1.2-2)). We also use the isomorphic vector spaces $(\mathbb{R}^N;\ +,-,\cdot)$ (cf. (3.1-8)f) and their computer realizations $(S\mathbb{R}^N;\ \widetilde{+},\widetilde{-},\widetilde{\cdot})$ (cf. [1], [11]). All operators $+,-,\cdot$ which occur in the right members of (7.1.4-4)-(7.1.4-6) to follow are such operations.

Now let $\overset{o}{h}, \hat{h} \in \mathbb{R}^N$, and let $h = (\overset{o}{h}, \hat{h})^T$. The *functoid* $(S_N(\mathcal{M}); \ S_N(\Omega))$ can be realized as follows.

$$p \boxplus q =: u \quad \text{with} \quad iu := (p_i + q_i)_{i=0}^N. \tag{7.1.4-4}$$

$$p \boxminus q =: u \quad \text{with} \quad iu := (p_i - q_i)_{i=0}^N. \tag{7.1.4-5}$$

$$p \boxdot q =: u \quad \text{with} \quad iu := \mathcal{A}(S_N)*h \tag{7.1.4-6}$$

$$= \overset{o}{h} + \overset{\wedge}{\mathcal{A}}(S_N)*\hat{h},$$

where

$$\overset{o}{h} = \left(\sum_{j=0}^i p_j \cdot q_{i-j} \right)_{i=0}^N \quad \text{and} \quad \hat{h} = \left(\sum_{j=0}^i p_j \cdot q_{i-j} \right)_{i=N+1}^{2N}$$

Integration $\boxplus$ is treated similarly to multiplication. We defer treatment of division, $\boxslash$, dealing first with the interval functoid and the operations $\Diamond$, $\Diamond$, and $\Diamond$.

Then let P, Q, and U be interval polynomials in $Isp \ \Phi_N = IS_N(\mathcal{M})$ (cf. (3.2)).

$$P(x) = \sum_{i=0}^N P_i x^i, \quad Q(x) = \sum_{i=0}^N Q_i x^i$$

and

$$U(x) = \sum_{i=0}^N U_i x^i = \Phi * iU \in IS_N(P\mathcal{M}).$$

Here we make use of the directed rounding matrices $\mathcal{A}(IS_N)$ (cf. (3.2.2-2b)) and $\overset{\wedge}{\mathcal{A}}(IS_N)$ (cf. (3.2.2-4)). We also use the isomorphic interval vector spaces $(I\mathbb{R}^N; \ +, -, \cdot)$ and their computer realizations $(IS(I\mathbb{R}^N); \ \tilde{+}, \tilde{-}, \tilde{\cdot})$ (cf. [1], [11]). All operations $+, -, \cdot$ which occur in the right members of (7.1.4-7)-(7.1.4-9) to follow are such interval operations.

Let $\overset{\circ}{H}, \hat{H} \in I\mathbb{R}^N$, and let $H = (\overset{\circ}{H}, \hat{H})^T$. The *interval functoid* $(IS_N(P\mathcal{M}); IS_N(\Omega))$ can be realized as follows.

$$P \,\diamondplus\, Q =: U \quad \text{with} \quad iU := (P_i + Q_i)_{i=0}^N. \tag{7.1.4-7}$$

$$P \,\diamondminus\, Q =: U \quad \text{with} \quad iU := (P_i - Q_i)_{i=0}^N. \tag{7.1.4-8}$$

$$P \,\diamondtimes\, Q =: U \quad \text{with} \quad iU := \mathcal{A}(IS_N)^*H \tag{7.1.4-9}$$

$$= \overset{\circ}{H} + \overset{\wedge}{\mathcal{A}}(IS_N)^*\hat{H},$$

where

$$\overset{\circ}{H} := \left(\sum_{j=0}^i P_j \cdot Q_{i-j} \right)_{i=0}^N \quad \text{and} \quad \hat{H} := \left(\sum_{j=0}^i P_j \cdot Q_{i-j} \right)_{i=N+1}^{2N}$$

Integration $\oint$ is treated similarly to multiplication.

Division

The implementation of division $\boxed{/}$ requires a more elaborate treatment. Indeed let $p(x)$ and $q(x)$ be elements of $sp\ \Phi_N$. The implementation of the quotient $p(x)/q(x)$ is composed of five steps.

step 1. approximate reciprocation

We compute an approximation $\widetilde{u}(x)$ to the reciprocal

$$u(x) = \frac{1}{q(x)}, \tag{7.1.4-10}$$

where

$$q(x) := \sum_{i=0}^N q_i x^i$$

and

$$\tilde{u}(x) := \sum_{i=0}^{N} u_i x^i.$$

The approximation is obtained by replacing (7.1.4-10) by

$$S_N(\tilde{u}(x) \cdot q(x)) = 1.$$

Then

$$u_0 := 1/q_0,$$

$$u_i := (\sum_{k=0}^{i-1} u_k \, q_{i-k})/q_0, \quad i = 1, 2, ..., N.$$

step 2. iterative improvement

The equation $vq - 1 = 0$ is replaced by the Newton-like iteration

$$v_{i+1} = v_i - \frac{v_i \cdot q - 1}{q}. \qquad (7.1.4-11)$$

To avoid the division here, $1/q$ is replaced by v_i itself. Thus we have

$$v_{i+1} = S_N v_i (2 - v_i \cdot q), \quad i = 0, 1, ...,$$

$$\qquad\qquad\qquad\qquad\qquad (7.1.4-12)$$

$$v_0 := \tilde{u}.$$

This iteration is convergent if $\| 1 - v_i \cdot q \| < 1$. The iteration is continued until a numerical convergence test is satisfied. For instance, until

$$\| v_{i+1} - v_i \| < \delta \, \| v_i \|,$$

when δ is some small prescribed tolerance ($\sim 10^{-10}$).

Then take

$$1 \; \boxed{/} \; q := v_{i+1} =: \tilde{v}. \qquad (7.1.4-13)$$

step 3. division

$$p \; \boxed{/} \; q \; = \; p \; \boxed{\cdot} \; \tilde{v} \qquad\qquad (7.1.4\text{-}14)$$

step 4. validation, inclusion

Let v be the fixed point of (7.1.4-12), which we write as

$$v = v - v(v{\cdot}q - 1).$$

Replace this equation by

$$v = v - \tilde{v}(v{\cdot}q - 1). \qquad\qquad (7.1.4\text{-}15)$$

Now we rearrange (7.1.4-15)

$$v = \tilde{v} - \tilde{v}(\tilde{v}{\cdot}q - 1) + (1 - \tilde{v}{\cdot}q)(v - \tilde{v}). \qquad (7.1.4\text{-}16)$$

Then the correction $\Delta v := v - \tilde{v}$ to $\tilde{v}$ is determined by the fixed point equation (7.1.4-16), viz.

$$\Delta v = \tilde{v}(1 - \tilde{v}{\cdot}q) + (1 - \tilde{v}{\cdot}q)\Delta v. \qquad\qquad (7.1.4\text{-}17)$$

We suppose that $\| 1 - \tilde{v}{\cdot}q \| < 1$. In this case (7.1.4-17) may be used to compose the following iterative computation in $(IS_N(\mathcal{M}); IS_N(\Omega))$:

$$U_0 := \Diamond(\tilde{v}{\cdot}(1 - \tilde{v}{\cdot}q))$$

repeat

$$U_{i+1} = U_0 \; \diamondsuit \; \Diamond(1 - \tilde{v}{\cdot}q) \; \diamondsuit \; U_i$$

until

$$U_{i+1} \subset U_i$$

The termination of this process implies that $\Diamond v \in U_{i+1}$ and

$$v = \frac{1}{q} \in \tilde{v} + U_{i+1} =: V.$$

step 5. validated quotient

Finally we have

$$p/q \in p \diamondsuit V.$$

Implementation of division, $\diamondsuit$ in the interval functoid proceeds by extension of the treatment for $\boxdot$.

Remark 7.1.4-1:

Recall the discussion in Section 3.2 concerning the symbol $\sum_{i=0}^{N} A_i \phi_i$ for which the interval ultra-arithmetic was defined, and for which the interval ultra-arithmetic is implemented in the present section. In that discussion we pointed out the dual nature of such symbols: first as a formal linear combination of intervals for each $\xi \in X$ and second as a set of functions. As a set of functions, these symbols are relevant for definition and implementation of interval ultra-arithmetic. However as such, the implementation of the interval ultra-arithmetic as described here, being so straightforward, raises a question.

The boundaries of a set $\sum_{i=0}^{N} A_i \phi_i$ are composed of piecewise polynomials. The boundaries of the set resulting from implementation of an ultra-arithmetic operation are then in general even more complicated piece-wise polynomials. Are such resultant sets contained in the indicated resultant sets, as so directly specified in this section.

To see that this is the case, we use the alternate characterization of the symbol $\sum_{i=0}^{N} A_i \phi_i$ as a graph. For each $\xi \in X$ an ultra-arithmetic operation amounts to an ordinary interval operation on intervals of reals, say

$$\sum_{i=0}^{N} A_i \phi_i(\xi), \quad \sum_{i=0}^{N} B_i \phi_i(\xi) \in I\mathbb{R}.$$

The operations implemented in this section clearly give a correct result for each $\xi \in X$. Then the implementation is correct for all of X.

7.2 ILLUSTRATIVE COMPUTATION

Here we treat computations illustrating the methodology of this monograph. The Taylor rounding and the Chebyshev rounding are both used.

In Section 7.2.1 we treat a general linear system of differential equations with boundary conditions. These computations are conducted in the isomorphic image space $\mathbb{R}^{N'}$ for approximation and in $\mathbb{IR}^{N'}$ for validation.

In Section 7.2.2 we treat a general nonlinear system by means of iteration directly in the functoid $(S_N(\mathcal{M}); \boxplus, \boxminus, \boxdot, \boxslash, \boxslash\!\!\!\!\boxslash)$ for approximation and in the functoid $(IS_N(\mathcal{M}); \diamondplus, \diamondminus, \diamonddot, \Diamond, \oiint)$ for validation.

In Section 7.2.3 we treat IRC.

7.2.1 Linear Differential Equations

We deal with linear equations in the two forms $\mathcal{L}(y) = 0$ and $y = \ell(y)$. The latter is needed for validation of a solution. $\mathcal{L}(y)$ is replaced by its isomorphic representation in $\mathbb{R}^{N'}$ corresponding to the finite part Φ_N of the basis of choice Φ of $\mathcal{M}$ as in Section 4.1:

$$\mathcal{A}(S_N)\mathcal{A}(\mathcal{L})v = 0 \qquad (7.2.1\text{-}1)$$

(cf. (4.1-3)-(4.1-4)).

For the isomorphic representation $\mathcal{A}(S_N)$ of the rounding operator S_N cf. (3.1.2-2), (3.1.2-3)f, and for an example cf. (3.1.2-34)f. For the isomorphic representation of $\mathcal{A}(\mathcal{L})$, we make use of the relation (4.1.2-2) and of operators such as $\mathcal{A}\left(\dfrac{d}{dx}\right)$, $\mathcal{A}\left(\int_0^x\right)$, $\mathcal{A}\left(\underset{x}{Sh}^{\alpha x+\beta}\right)$, ..., cf. (4.1.2-3)-(4.1.2-12).

For validation, the equation $y = \ell(y)$ is replaced by its isomorphic representation in $\mathbb{IR}^{N'}$ corresponding to the $\mathbb{I}\!\!\mathit{sp}\Phi_N$ of $\boldsymbol{P\mathcal{M}}$. (For a discussion of the latter notion, see the introductory remarks of Section

3.2.). Corresponding to (4.1.2-19), this representation is

$$W := W_0 + IR_N(\ell)*W, \qquad (7.2.1-2)$$

(For an example cf. (4.1.2-18)-(4.1.2-20)).

In the following numerical examples we have chosen $\mathcal{M} = L^2[-1, 1]$, the basis

$$\Phi_{12} = \{1, x, ..., x^{12}\}$$

and the Chebyshev rounding (cf. Section 3.1.2(i)).

All of our examples are developed as special cases of a general boundary value problem, which we now describe.

Let $a(x)$, $b(x)$, $c(x)$, and $d(x)$ be polynomials in $sp\{\Phi_{12}\}$, and let r_i, g_{ij}, h_{ik}, ξ_{ij}, $\eta_{ik} \in \mathbb{R}$, $j = 1, ..., q_i$, $k = 1, ..., q'_i$, $i = 1, 2$. The boundary-value problem is

$$a(x)\frac{d^2}{dx^2}y(x) + b(x)\frac{d}{dx}y(x) + c(x)y(x) = d(x), \quad -1 < x < 1$$
$$\qquad (7.2.1-3)$$
$$c(y) := \sum_{j=1}^{q_i} g_{ij}y(\xi_{ij}) + \sum_{k=1}^{q'_i} h_{ik}\frac{d}{dx}y(\eta_{ik}) = r_i, \quad i = 1, 2.$$

Referring to Section 3.1.2(iii), we see that for the isomorphic correspondent in $\mathbb{R}^{13}$ of this boundary value problem, namely, for $\mathcal{A}(S_{11})$ $\mathcal{A}(\mathcal{L})v = 0$, we have

$$\begin{cases} \mathcal{A}(S_{11})(\mathcal{A}(\ell)*v - id) = 0, \\ \mathcal{A}(c)*v - r = 0. \end{cases} \qquad (7.2.1-4)$$

Here

$$\ell y(x) := \left(a(x)\frac{d^2}{dx^2} + b(x)\frac{d}{dx} + c(x) \right)y(x),$$

and

$$id := (d_0, ..., d_{10}),$$

the latter is the vector that is the isomorphic image of the polynomial

$$d(x) = \sum_{i=0}^{12} d_i x^i.$$

(Recall that degrees of freedom in $d(x)$ are dropped upon passage to id in order to accommodate the constraints.)

$\mathscr{A}(c)$ is a 2×13 matrix, where the in^{th} element is

$$(\mathscr{A}(c))_{i,n} := \sum_{j=1}^{q_i} g_{ij}\xi_{ij}^n + n\sum_{k=1}^{q'_i} h_{ik}\eta_{ik}^{n-1}, \quad i = 1,2; \quad n = 0,1,...,12.$$

Here $\eta_{ik}^{-1} := 0$.

Theorem 4 of Chapter 2, which we use for the validation process, requires a compact operator ℓ. For this reason, the problem (7.2.1-3) is transformed into an integral equation through the transformation

$$u := y''.$$

Then for the equation $u = \ell u$, the new operator ℓ is given as

$$\ell u := \frac{d}{a} - \frac{b}{a}\left(\beta + \int_0^x u\,dx\right) - \frac{c}{a}\left(\alpha + \beta x + \int_0^x \int_0^x u\,dx^2\right), \quad (7.2.1\text{-}5)$$

where α and β are determined by

$$c\left(\alpha + \beta x + \int_0^x \int_0^x u\,dx^2\right) = r. \qquad (7.2.1\text{-}6)$$

The equation $u = \ell u$ may be transformed into an equation of the form (7.2.1-2) from which a validation of u can be obtained, say, $u \in U$. Then y, y', and y'' are validated as well, and, in particular, we have

$$y'' \in U,$$

$$y' \in \beta + \int_0^x U\,dx,$$

$$y \in \alpha + \beta x + \int_0^x \int_0^x U\,dx^2 =: Y(x).$$

The computed inclusion $Y(x)$ for the solution $y(x)$ of each problem is given by

$$Y(x) = \sum_{i=0}^{12} Y_i x^i. \qquad (7.2.1\text{-}7)$$

We stress that in addition to the *bound represented by the inclusion* $y(x) \in Y(x)$, $x \in [-1, 1]$, *the computations validate the existence and uniqueness of the solution* $y(x)$ *of each particular example.*

In the examples of the linear case that follow we specify the differential equation and the boundary conditions as in (7.2.1-4) and the exact solution, when the last is known. The coefficients Y_i, $i = 0, ..., 12$, of the solution inclusion (cf. (7.2.1-3)) are then listed, followed by a listing of values of $Y(x)$ itself at a number of points.

We employ a compact form for displaying intervals. For example, the interval

$$[1.234E - 20, \quad 1.275E - 20]$$

is written simply

$$1.2^{75}_{34}E - 20.$$

The examples now follow:

1. Boundary-value problem

$$(32 + 2x)y'' - (13 + x)y' - y = 0$$
$$7y(-1) + 8y(1) - 15y'(-1) - 17y'(1) = 0$$
$$y(0) + 32y'(0) = 4$$

Exact solution

$$y(x) = e^{x/2}/\sqrt{x + 16}$$

digit position	**1.23**	**456**	**789**	**012**	
Y_0	= 2.50	000	000	00_0^1	$E-01$
Y_1	= 1.17	187	$\frac{500}{499}$	$\frac{000}{999}$	$E-01$
Y_2	= 2.77	099	609	37_4^6	$E-02$
Y_3	= 4.39	580	281	57_5^6	$E-03$
Y_4	= 5.25	563	955	30_7^8	$E-04$
Y_5	= 5.04	671	906	38_1^2	$E-05$
Y_6	= 4.05	156	202	29_8^9	$E-06$
Y_7	= 2.79	561	963	02_5^6	$E-07$
Y_8	= 1.69	184	540	88_3^4	$E-08$
Y_9	= 9.11	938	707	31_8^9	$E-10$
Y_{10}	= 4.43	195	219	85_0^1	$E-11$
Y_{11}	= 1.97	292	636	69_3^4	$E-12$
Y_{12}	= 8.01	910	182	99_0^2	$E-14$
$Y(-1.0) =$	1.56	605	542	93_5^8	$E-01$
$Y(-0.5) =$	1.97	815	596	71_5^8	$E-01$
$Y(0.0)$	= 2.50	000	000	00_0^1	$E-01$
$Y(0.5)$	= 3.16	105	205	99_0^2	$E-01$
$Y(1.0)$	= 3.99	873	643	89_2^8	$E-01$

2. Boundary-value problem

$$4y'' - y = 0.5 + 0.5x$$

$$y(-1) = 0$$

$$y(1) = 1$$

Exact solution

$$y(x) = \frac{2}{\sinh 1} \sinh \frac{x+1}{2} - \frac{x+1}{2}$$

digit position	1.23	456	788	012	
Y_0	= 3.86	818	883	9^{71}_{69}	$E-01$
Y_1	= 4.59	517	375	6^{68}_{67}	$E-01$
Y_2	= 1.10	852	360	4^{97}_{96}	$E-01$
Y_3	= 3.99	798	906	5^{29}_{28}	$E-02$
Y_4	= 2.30	942	417	7^{01}_{00}	$E-03$
Y_5	= 4.99	748	633	0^{98}_{96}	$E-04$
Y_6	= 1.92	452	014	7^{36}_{34}	$E-05$
Y_7	= 2.97	469	441	3^{80}_{78}	$E-06$
Y_8	= 8.59	160	815	5^{82}_{79}	$E-08$
Y_9	= 1.03	285	744	9^{38}_{37}	$E-08$
Y_{10}	= 2.38	651	520	4^{72}_{71}	$E-10$
Y_{11}	= 2.36	216	683	6^{77}_{76}	$E-11$
Y_{12}	= 4.54	574	324	7^{08}_{06}	$E-13$

$$Y(-1.0) = \pm 2.2 \qquad\qquad\qquad E-12$$

$$Y(-0.4) = 2.18 \; 243 \; 676 \; 22^{4}_{0} \;\; E-01$$

$$Y(0.0) = 3.86 \; 818 \; 883 \; 9^{71}_{69} \;\; E-01$$

$$Y(0.4) = 5.90 \; 985 \; 247 \; 36^{6}_{2} \;\; E-01$$

$$Y(1.0) = \begin{matrix} 1.00 \\ .99 \end{matrix} \begin{matrix} 000 \\ 999 \end{matrix} \begin{matrix} 000 \\ 999 \end{matrix} \begin{matrix} 001 \\ 9998 \end{matrix} \; E00$$

3. Boundary-value problem

$$(400 + 4x)y'' - (100 + x)y = -400 + 396x + 104x^2 + x^3$$

$$y(-1) + y'(-1) = 9$$

$$y(0) - 2y'(0) = 4$$

Exact solution

$$y(x) = 8e^{\frac{x+1}{2}} - (x + 2)^2$$

digit position	**1.23**	**456**	**789**	**012**	
Y_0	= 9.18	977	016	5^{62}_{59}	$E + 00$
Y_1	= 2.59	488	508	2^{81}_{79}	$E + 00$
Y_2	= 6.48	721	270	$^{703}_{698}$	$E - 01$
Y_3	= 2.74	786	878	4^{51}_{49}	$E - 01$
Y_4	= 3.43	483	598	06^{4}_{0}	$E - 02$
Y_5	= 3.43	483	598	0^{20}_{18}	$E - 03$
Y_6	= 2.86	236	331	69^{4}_{1}	$E - 04$
Y_7	= 2.04	454	534	25^{8}_{7}	$E - 05$
Y_8	= 1.27	784	082	32^{9}_{6}	$E - 06$
Y_9	= 7.09	896	074	7^{41}_{38}	$E - 08$
Y_{10}	= 3.54	949	163	2^{70}_{66}	$E - 09$
Y_{11}	= 1.62	354	731	75^{4}_{3}	$E - 10$
Y_{12}	= 6.76	170	958	0^{33}_{26}	$E - 12$
$Y(-1.0) =$	$^{6.99}_{7.00}$	$^{999}_{000}$	$^{997}_{004}$	$^{997}_{004}$	$E + 00$
$Y(-0.5) =$	8.02	220	333	3^{53}_{48}	$E + 00$
$Y(0.0)$	= 9.18	977	016	5^{62}_{59}	$E + 00$
$Y(0.5)$	= 1.06	860	001	3^{30}_{28}	$E + 01$
$Y(1.0)$	= 1.27	462	546	27^{8}_{6}	$E + 01$

4. Boundary-value problem

$$(400.13 + 0.39x)y'' - (98.07 + x)y = -399 + 396x + 104x^2 + 1.07x^3$$

$$y(-1) + 3y(-0.7) - 2y(0.7) + y'(-1) = 9$$

$$y(0) - 2y'(0) + y(-0.5) = 12$$

digit position		1.23	456	789	012	
Y_0	=	9.12	903	964	08_8^9	$E + 00$
Y_1	=	2.55	243	232	93_1^2	$E + 00$
Y_2	=	6.20	154	596	73_3^4	$E - 01$
Y_3	=	2.72	812	318	32_6^7	$E - 01$
Y_4	=	3.47	246	754	81_7^8	$E - 02$
Y_5	=	3.53	414	479	89_6^7	$E - 03$
Y_6	=	3.04	125	735	21_0^1	$E - 04$
Y_7	=	2.24	784	097	56_5^6	$E - 05$
Y_8	=	1.47	235	904	60_1^2	$E - 06$
Y_9	=	8.59	566	377	36_8^9	$E - 08$
Y_{10}	=	4.56	670	855	98_2^3	$E - 09$
Y_{11}	=	2.22	953	820	03_2^3	$E - 10$
Y_{12}	=	9.99	406	714	1_{67}^{72}	$E - 12$

$Y(-1.0) =$		6.95	542	315	87_3^6	$E + 00$
$Y(-0.5) =$		7.97	582	501	77_4^7	$E + 00$
$Y(0.0)$	=	9.12	903	964	08_8^9	$E + 00$
$Y(0.5)$	=	1.05	966	816	62_2^3	$E + 01$
$Y(1.0)$	=	1.26	130	258	72_7^9	$E + 01$

5. Boundary-value problem

$$(1 + 0.1x + 0.005x^2 + 0.0001666x^3)y'' + (x - 0.3334x^3 + 0.0416x^5)y'$$
$$+ \frac{1}{5}(1 + x + x^2 + x^3 + x^4)y = 1 - 0.25x^2 + 0.015625x^4$$
$$- 0.00043403x^6 + 0.00000678168x^8$$

$$y(-1) + 0.7y'(0.3) + 10y(1) = 1$$
$$11y'(-1) + 0.3y(0.3) + 0.1y'(1) = 0$$

digit position　　　**1.2345**

Y_0 $= -\ 1.715^3_7$ $E + 00$

Y_1 $=\ \ \ 1.602^5_2$ $E + 00$

Y_2 $=\ \ \ 6.71^{60}_{49}$ $E - 01$

Y_3 $= -\ 2.856^1_6$ $E - 01$

Y_4 $= -\ 1.2843^2_4$ $E - 01$

Y_5 $=\ \ \ 7.46^9_7$ $E - 02$

Y_6 $=\ \ \ 2.682^6_2$ $E - 02$

Y_7 $= -\ 2.78^{36}_{41}$ $E - 02$

Y_8 $= -\ 6.06^{85}_{93}$ $E - 03$

Y_9 $=\ \ \ 5.3^{906}_{897}$ $E - 03$

Y_{10} $=\ \ \ 9.25^{96}_{85}$ $E - 04$

Y_{11} $= -\ 6.52^{33}_{43}$ $E - 04$

Y_{12} $= -\ 7.86^{36}_{45}$ $E - 05$

$Y(-1.0) = -\ 2.51^{86}_{94}$ $E + 00$

$Y(-0.5) = -\ 2.32^{26}_{31}$ $E + 00$

$Y(0.0)\ \ \ = -\ 1.715^3_7$ $E + 00$

$Y(0.5)\ \ \ = -\ 7.87^3_9$ $E - 01$

$Y(1.0)\ \ \ =\ \ \ 2.1^{81}_{73}$ $E - 01$

7.2.2 Nonlinear Differential Equations

The class of problem types for a system of nonlinear differential equations is so wide that we refrain from specifying the detailed form of the differential equation as in the linear case just treated in Section 7.2.1. Thus we take our problem to be of the form

$$y' = f(x,y,y'), \quad x \in [-1, 1], \tag{7.2.2-1a}$$

with the boundary condition

$$R(y) = r. \tag{7.2.2-1b}$$

Here y and f, R and r, are n vectors. Each component of y is an element of $\mathcal{M} := L^2[-1, 1]$, while each component of r is an element of $\mathbb{R}$.

The problem (7.2.2-1) is expressed in terms of a compact mapping as required by the fixed point theorems of Chapter 2, namely, in the form

$$z = f(x, a + \int_0^x z\,dx, z), \tag{7.2.2-2a}$$

where the n vector

$$a := a(r, z) \tag{7.2.2-2b}$$

is determined by the boundary conditions (7.2.2-1b), viz.,

$$R(a + \int_0^x z\,dx) = r. \tag{7.2.2-3}$$

To apply Theorem 4, we assume that $f(x,u,v)$ is square integrable in each argument, continuous in u, and contractive in v.

In Section 4.2.2 we saw that a wide class of computations are obtained from differential equations composed of rational operations in the functoid $(\mathcal{M}; +, -, \bullet, /, \int)$ (compare (4.2.2-1)-(4.2.2-6)). Thus for our computational purposes here, we restrict the boundary-value problem to such a case. As before, we indicate this by adjoining the set of operators $\Omega = \{+, -, \bullet, /, \int\}$ to the argument list of the functions defining the problem. Thus (7.2.2-2) is rewritten

(a) $\quad v = f(x, a + \int_0^x v\,dx, v; \Omega),$

$$(7.2.2\text{-}4)$$

(b) $\quad a = a(r, v; \Omega).$

The boundary condition (7.2.2-4b) here involves values of functions at points, so that the operations Ω appearing there are in reality the restrictions of Ω to ordinary arithmetic in $\mathbb{R}$.

Our computation consists of iterating (7.2.2-4) directly in the functoid $(S_N(\mathcal{M}); \, S_N(\Omega)) = (S_N(\mathcal{M}); \, \boxplus, \, \boxminus, \, \boxdot, \, \boxslash, \, \oint)$ (cf. Section 3.1). (This is in contrast to the linear case treated in Section 7.2.1, where the problem was cast into its isomorphic image in $\mathbb{R}^{N'}$ (cf. (7.2.1-1)).) To perform the iteration, we replace Ω in (7.2.2-4) by $S_N(\Omega)$, and we proceed to the following program, which uses some prescribed norm and some prescribed error tolerance ε:

$$v_0 := v_0$$

repeat

$$a := a(r, v_i; \, S_N(\Omega))$$

$$v_{i+1} := f(x, a \boxplus \oint_0^x v_i\,dx, v_i; \, S_N(\Omega)) \qquad (7.2.2\text{-}5)$$

until

$$\text{norm } (v_{i+1} - v_i) \leq \varepsilon \cdot \text{norm } (v_i)$$

Convergence of this process will deliver an approximation $\tilde{v} \sim v$ of the solution of (7.2.2-4), or equivalently an approximation $\tilde{v} \sim y'$ of the derivative of the solution of (7.2.2-1). The approximation to y itself is given by

$$y \cong a(r, \tilde{v}) \boxplus \oint_0^x \tilde{v}\,dx. \qquad (7.2.2\text{-}6)$$

validation

The validation step proceeds in terms of the isotonic computer-screen implementation

$$f(x,U,V; IS_N(\Omega)): \quad (IS_N(P\mathcal{M}))^2 \rightarrow IS_N(P\mathcal{M}),$$

which is obtained from f by replacing the operations Ω by the corresponding operations $IS_N(\Omega)$ in the interval functoid $(IS_N(P\mathcal{M}); IS_N(\Omega)) = (IS_N(P\mathcal{M}); \lozenge, \lozenge, \lozenge, \lozenge, \oint)$ (cf. (3.2-5)).

The validation process then consists of the following program:

$$V_0 := \tilde{v}$$

repeat

$$A := a(r, V_i; IS_N(\Omega)) \qquad (7.2.2-7)$$

$$V_{i+1} := f(x, A \lozenge \oint_0^x V_i dx, V_i; IS_N(\Omega))$$

until

$$V_{i+1} \subset V_i$$

Then $V := V_{i+1}$ provides an inclusion of $y'(x)$, the derivative of the solution of (7.2.2-1), i.e., $y'(x) \in V$. The inclusion of y itself is given by

$$y \in a(r,V;IS_N(\Omega)) + \int_0^x V(x)dx =: Y(x).$$

Conceivably the termination condition $V_{i+1} \subset V_i$ in (7.2.2-7) may not be achieved within a specified number of iterations. This may be due to:

(i) poor condition of the numerical problem,

(ii) the problem being solved has no solution in the class in question.

In the case that the iteration $V_{i+1} := f$ in (7.2.2-7) is replaced by a locally linearized form as in (7.1.3-2) in order to establish uniqueness. Then failure to achieve the convergence criterion $V_{i+1} \overset{o}{\subset} V_i$ may signal a lack of local uniqueness of the solution of the original problem.

The examples now follow. As in the linear case we use the polynomial basis (cf. (7.2.1-3)). The basic domain is $X = [-1, 1]$ and $N = 10$. Two roundings S_N are variously used: the Taylor rounding and the Chebyshev rounding.

The inclusion computed for $y(x) = (y_j(x))$ is denoted by $Y(x) = Y_j(x)$, where

$$Y_j(x) = \sum_{i=0}^{10} Y_{ji}x^i \qquad (7.2.2\text{-}8)$$

for each component: $j = 1$ or $j = 1,2$ as the case may be. We stress that in addition to the *bound represented by the inclusion* $y(x) \in Y(x)$, $x \in [-1, 1]$, *the computations validate the existence of the solution* $y(x)$ *for each praticular example.*

For simplicity in the case where we have two functions, viz., $j = 1,2$ as in example 2 below, we denote both $Y_1(x)$ and $Y_2(x)$ by $Y(x)$, and we label the corresponding columns by the component names y_1 and y_2.

1. The initial-value problem is

$$y'(t) = -\frac{1}{2} - (\frac{1}{8}t^2 + \frac{1}{4}t + \frac{1}{8})y^2(t),$$
$$y(-1) = 1.$$

Then the equations of (7.2.2-4) on which the iterations are based become

(a) $\quad z = -\dfrac{1}{2} - (\dfrac{1}{8}x^2 + \dfrac{1}{4}x + \dfrac{1}{8})(a + \displaystyle\int_0^x zdt)^2$

(b) $\quad a = 1 - \displaystyle\int_0^{-1} zdt.$

This initial-value problem has a zero close to but less than unity. A validation of this zero $t_0 \in T$ is obtained by applying Newton's method for x near unity to the two polynomials $p(x)$ and $q(x)$ that specify the inclusion $Y(x) = [p(x),q(x)]$. The tabulated results show that the Chebyshev rounding is superior.

	Taylor rounding		Chebyshev rounding			
digit position	**1.2345**		**1.23**	**456**	**789**	
Y_0	$= 4.8^{40}_{36}$	$E-01$	4.83	79^{3}_{2}	$^{2}_{8}$	$E-01$
Y_1	$= -5.29^{1}_{4}$	$E-01$	-5.29	25^{6}_{7}	$^{8}_{1}$	$E-01$
Y_2	$= 2.7^{8}_{3}$	$E-03$	2.74	91^{7}_{1}		$E-03$
Y_3	$= 2.11^{6}_{2}$	$E-02$	2.11	401	$^{9}_{5}$	$E-02$
Y_4	$= -2.2^{0}_{4}$	$E-03$	-2.21	58^{9}_{6}		$E-03$
Y_5	$= -7.3^{28}_{39}$	$E-03$	-7.33	10^{4}_{6}		$E-03$
Y_6	$= 7.5^{4}_{1}$	$E-04$	7.42	75^{7}_{2}		$E-04$
Y_7	$= 4.4^{5}_{2}$	$E-04$	4.37	04^{5}_{2}		$E-04$
Y_8	$= -1.9^{83}_{92}$	$E-04$	-1.81	47^{6}_{9}		$E-04$
Y_9	$= -9.6^{9}_{5}$	$E-05$	-8.84	2^{23}_{33}		$E-05$
Y_{10}	$= 2.^{82}_{79}$	$E-05$	1.32	56^{8}_{4}		$E-05$
$Y(-1.0) =$	$^{1.0004}_{9.99965}$	$E\ 00$	$^{1.00}_{9.99}$	$^{000}_{999}$	$^{04}_{962}$	$E^{\ 00}_{-01}$
$Y(-0.5) =$	7.46^{8}_{3}	$E-01$	7.46	564	$^{8}_{2}$	$E-01$
$Y(0.0) =$	4.84^{40}_{36}	$E-01$	4.83	79^{3}_{2}	$^{2}_{8}$	$E-01$
$Y(0.5) =$	2.2^{24}_{18}	$E-01$	2.22	14^{1}_{0}	$^{2}_{6}$	$E-01$
$Y(1.0) =$	$-^{2.98}_{3.06}$	$E-02$	-3.01	98^{0}_{8}		$E-02$
T	$= 9.^{4}_{3}$	$E-01$	9.39	62^{2}_{0}		$E-01$

2. The boundary-value problem is

$$u' = 0.1uv^2 + 0.1,$$
$$v' = 0.1u^2 + 0.5x + 0.05,$$

$$u(-1) = 1, \quad v(1) = 0.$$

The form corresponding to (7.2.2-4) used for iteration is

$$y_1 = 1.1 - 0.1\int_0^{-1} y_1 y_2^2 dt + 0.1x + 0.1\int_0^x y_1 y_2^2 dt,$$

$$y_2 = -0.075 + 0.05x + 0.025x^2 + 0.1\int_0^x 0.1 y_1^2 dt.$$

With Taylor rounding we get

digit position	y_1 1.23	456	789	y_2 1.23	456	789
Y_0	$= 1.10$	859	689^3_0 E+00	-2.09	629	54^5_7 E-01
Y_1	$= 1.04$	871	6^{80}_{78} E-01	1.72	898	70^{71}_{65} E-01
Y_2	$= -3.78$	764	6^6_9 E-03	3.66	260	41^8_6 E-02
Y_3	$= 2.78$	281	1^7_1 E-04	8.66	707	$^6_2$ E-05
Y_4	$= 3.95$	293	5^5_3 E-04	-4.4	576	$^6_1$ E-06
Y_5	$= 5.57$	795	0^7_0 E-05	1.89	831	2^6_4 E-05
Y_6	$= 1.17$	572	$^3_1$ E-06	3.40	793	$^{52}_{48}$ E-06
Y_7	$= 7.25$	64^6_2	E-08	1.62	702	$^{33}_{25}$ E-07
Y_8	$= 1.17$	398	$^{48}_{32}$ E-07	2.56	18^7_5	E-09
Y_9	$= 2.05$	437	$^4_1$ E-08	5.04	344	$^8_9$ E-09
Y_{10}	$= 1.20$	1^{60}_{59}	$^0_5$ E-09	1.14	376	$^{21}_{13}$ E-09

$Y(-1.0) =$	$\frac{1.00}{0.99}$	$\frac{000}{999}$	$\frac{0002}{9984}$ E+00	-3.46	009	05^5_8 E-01
$Y(-0.5) =$	1.05 .	520	233^8_4 E+00	-2.88	934	04^0_2 E-01
$Y(0.0) =$	1.10	859	689^3_0 E+00	-2.09	629	54^5_7 E-01
$Y(0.5) =$	1.16	014	707^4_1 E+00	-1.14	012	47^7_9 E-01
$Y(1.0) =$	1.21	041	166^7_3 E+00	± 9.2		E-10

With Chebyshev rounding we get

digit position	y_1 1.23	456	789	012		y_2 1.23	456	789	01	
Y_0 =	1.10	859	689	14^{8}_{3} $E+00$		-2.09	629	546	0^{6}_{9} $E-01$	
Y_1 =	1.04	871	678	77^{8}_{2} $E-01$		1.72	898	706	7^{9}_{6} $E-01$	
Y_2 =	-3.78	764	67^{4}_{5}	9^{2}_{0} $E-03$		3.66	260	417	$^{11}_{08}$ $E-02$	
Y_3 =	2.78	281	12^{3}_{2}	0^{2} $E-04$		8.66	707	18^{6}_{2}	$E-05$	
Y_4 =	3.95	293	516	5^{2} $E-04$		-4.43	576	6^{74}_{81}	$E-06$	
Y_5 =	5.57	796	056	7^{7}_{0} $E-05$		1.89	832	415	$^{5}_{3}$ $E-05$	
Y_6 =	1.17	581	20^{10}_{7}	$E-06$		3.40	794	78^{5}_{4}	$^{5}_{9}$ $E-06$	
Y_7 =	7.23	361	8^{8}_{3}	$E-08$		1.62	438	01^{7}_{6}	$^{7}_{8}$ $E-07$	
Y_8 =	1.17	245	55^{9}_{7}	$E-07$		2.54	041	$^{32}_{29}$	$E-09$	
Y_9 =	2.07	599	81^{7}_{3}	$E-08$		5.30	518	4^{93}_{88}	$E-09$	
Y_{10} =	1.32	235	5^{94}_{87}	$E-09$		1.16	002	90^{7}_{6}	$E-09$	

$Y(-1.0)$ =	$^{1.00}_{0.99}$	$^{000}_{999}$	$^{000}_{999}$	$^{004}_{971}$ $E\ 00$		-3.46	009	056	9^{4}_{8} $E-01$	
$Y(-0.5)$ =	1.05	520	233	6^{23}_{17} $E\ 00$		-2.86	934	041	3^{3}_{7} $E-01$	
$Y(0.0)$ =	1.10	859	689	14^{8}_{3} $E\ 00$		-2.09	629	546	0^{6}_{9} $E-01$	
$Y(0.5)$ =	1.16	014	707	2^{72}_{66} $E\ 00$		-1.14	012	477	$^{87}_{90}$ $E-01$	
$Y(1.0)$ =	1.21	041	166	5^{23}_{17} $E\ 00$		±1.9			$E-11$	

3. The boundary-value problem

$$u' = u^2 - 2xu + x^2 + 1,$$
$$au(0) + (a + 1)u(-1) + (a - 1)u(1) = 1, \qquad (7.2.2\text{-}9)$$

with a parameter a is treated next. The exact solution,

$$u(x) = \frac{1}{a - x} + x,$$

has a pole at $x = a$.

For iteration, the boundary-value problem (7.2.-9) is converted into the following form, corresponding to (7.2.2-4):

$$y(x) = \frac{1}{3a} \left(1 - (a+1) \int_0^{-1} w(x)dx - (a-1) \int_0^1 w(x)dx \right)$$
$$+ \int_0^x w(x)dx,$$

(7.2.2-10)

where $w(x) := y^2(x) - 2xy(x) + x^2 + 1$.

The computation is performed for various values of $a \in (1, 10]$. Let $\#Y(x)$ be the number of correct digits of the inclusion $Y(x)$ at x, and let the average number of correct digits be defined as follows

$$\overline{\#Y(x)} = \frac{1}{21} \sum_{i=-10}^{10} \#Y\left(\frac{i}{10}\right).$$

(7.2.2-11)

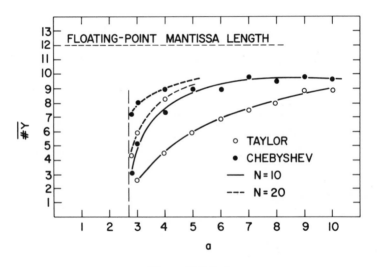

Figure 7.2.2-1

Figure 7.2.2-1 contains plots of $\overline{\#Y(x)}$ versus a for $N = 10, 20$ and the Taylor and Chebyshev roundings. The floating-point mantissa of the computer used is 12, this being indicated by the dashed line in the figure.

The only surprising feature of these plots is that they show that we could not obtain validation for a smaller than about 2.75. We had expected this feature to occur for a near unity, the location of the pole in the exact solution. A possible explanation for the barrier at 2.75 is that other solutions of the differential equation (possibly complex) have a pole of magnitude ~2.75. Such solutions may be supplying parasitic effects influencing our computation.

4. The initial-value problem

$$u' = 2^{-10}(-u + v - uv), \quad u(0) = 1,$$
$$v' = 2^{-10}\alpha(u - v - uv), \quad v(0) = 0,$$

is treated next with the independent variable $x \in [-1, 1]$. We perform the computation for $N = 10$ and $N = 20$. These small values of N (to which we are currently constrained by the computer in use) dictate the choice of the scaling 2^{-10}. In Fig. 7.2.2-2 we plot the average number of correct digits $\overline{\#U(x)}$ and $\overline{\#V(x)}$ (cf. (7.2.2-11)) of the inclusions $Y_1 = U$ and $Y_2 = V$ versus α for α in the interval $[2^6, 2^{12}]$. The plot is labeled # versus α with the curves for U and V labeled separately. Only the Chebyshev rounding is used. The floating-point mantissa length of the computer in use is 12 and is labeled by a dashed line in the figure.

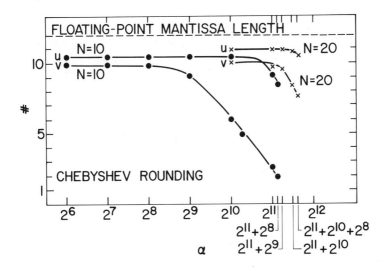

Figure 7.2.2-2

5. The following system of integral equations is treated next.

$$y_1 = 1 + \beta \int_1^x (y_1^3 y_2^3 + xy_1)dx,$$

$$y_2 = \alpha \left(2 + \int_{-1}^1 y_2 y_1^2 dx + \int_0^x y_2^2 y_1 dx \right).$$

The average number of correct digits in the inclusion (cf. (7.2.2-11)) is given in Tables 7.2.2-3 and 7.2.2-4 in which the values $N = 10, 20, 30$ along parameter values are variously used. Results for both Taylor and Chebyshev roundings are displayed.

Degree of polynomials	Taylor rounding	Chebyshev rounding
$N = 10$	no inclusion	3
$N = 20$	3.9	7.8
$N = 30$	7	9

$\alpha = 0.2, \beta = 0.4$

Table 7.2.2-3: Average number of correct digits.

Parameter		Taylor rounding	Chebyshev rounding
α	β		
0.1	0.3	4	7.5
	0.4	3.7	6.5
0.2	0.3	2	4.3
	0.4	no inclusion	3

$N = 10$

Table 7.2.2-4: Average number of correct digits.

7.2.3 IRC

Computations that illustrate the IRC process in both the linear and the nonlinear cases are presented in this section. Our framework is that of Section 5.4 (cf. (5.4-10)f). Let

$$\gamma_n = (5n, 5n + 1, ..., 5n + 10), \quad n = 0, 1, 2, ... \qquad (7.2.3\text{-}1)$$

The sequence Γ (cf. (5.4-4)) is taken to be

$$\Gamma = (\gamma_0, \gamma_0, \gamma_1, ..., \gamma_7, \gamma_8, \gamma_7, ..., \gamma_1, \gamma_0, \gamma_1, ...). \qquad (7.2.3\text{-}2)$$

Thus after the initial term γ_0, Γ is periodic with period $p = 16$. The equations in (5.4-10) and (5.4-12) are solved by iteration. In particular, since (5.4-1) has the form $y = ky := y + \mathscr{L}y = y + \ell y + \mathscr{L}0$, the corresponding iterations are

$$v_0^{j+1} = S_{N_0}^{\gamma_0}(kv_0^j) = S_{N_0}^{\gamma_0}(v_0^j + \ell v_0^j), \quad j = 0, 1, ... , \qquad (7.2.3\text{-}3)$$

with $v_0^0 = 0$ and

$$v_{n+1}^{j+1} = S_{N_{n+1}}^{\gamma_{n+1}}(v_{n+1}^j + \ell v_{n+1}^j - r_{n+1}), \quad j = 0, 1, ... , \qquad (7.2.3\text{-}4)$$

with $v_{n+1}^0 = 0$, respectively. Iteration is stopped when two successive iterates agree to all 12 places of the floating-point mantissa length employed. Likewise the IRC process, i.e., (7.2.3-2), is repeated ($n = 0,1,2,...$) until such agreement also occurs. Let us call $\bar{n}$ (resp. $j(\bar{n})$) the values of the indices n (resp. $j(n)$) in (7.2.3-3) (resp. (7.2.3-4)) at which these processes, as described, stop. Let

$$v^F = v_0^{j(0)}$$

and

$$v^L = \sum_{n=0}^{\bar{n}} v_n^{j(\bar{n})}$$

(cf. (5.4-13)). Briefly, v^F and v^L are the first and last functions resp. delivered by this computational implementation of IRC.

LINEAR EXAMPLES

1. Consider the boundary-value problem

$$y' = \zeta(x)y,$$

$$ay(0) + (a + 1)y(-1) - (a - 1)y(1) = 1,$$

where $\zeta(x) := 1/(a - x)$ with $a > 1$ a parameter. The exact solution

$$y = \frac{1}{a - x}$$

has a singularity at $x = a$.

In conformity with (7.2.2-2a) the boundary-value problem is transformed into the following integral equation:

$$y = ky := \frac{1}{a+2}(1 - (a + 1)\int_0^{-1} \zeta(x)ydx$$

$$+ (a - 1)\int_0^1 \zeta(x)ydx + \int_0^x \zeta(x)ydx.$$

For simplicity, we avoid division by a polynomial; hence we arrive at the form (7.2.2-5) of the integral equation by the a priori replacement of $\zeta(x)$ by $\tilde{\zeta}(x)$, where

$$\tilde{\zeta} := S_{50}\frac{1}{a-x} = \frac{1}{a}\sum_{j=0}^{50} \left(\frac{x}{a}\right)^j.$$

Thus we find

$$kv = v + \mathscr{L}v$$

$$= \frac{1}{a+2}\left(1 - (a + 1)\int_0^{-1}\tilde{\zeta}vdx + (a - 1)\int_0^1 \tilde{\zeta}vdx\right) + \int_0^x \tilde{\zeta}vdx,$$

so that

$$\ell v = \int_0^x \tilde{\zeta}(t) v \, dt.$$

The residue $r(v)$ evaluated at v is simply $\mathscr{L}v$.

In Fig. 7.2.3-1 we plot $\overline{\#v}$, the average number of correct digits (cf. (7.2.2-11)) in the solution versus a for the functions v^F and v^L.

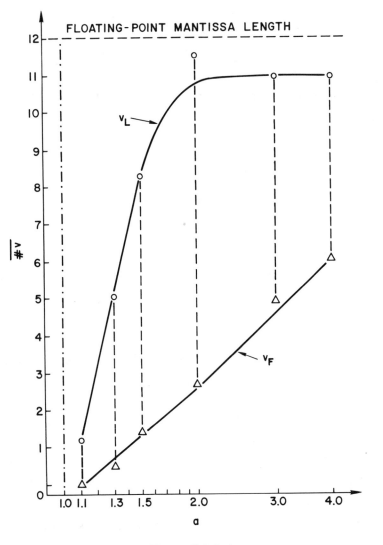

Figure 7.2.3-1

Now let v^Q be the function produced by this implementation of the IRC process after Q periods in the sequence Γ. (Thus, for example, $v^{Q=0} = v^F$.) Let

$$v^Q = \sum_{j=0}^{50} v_j^Q x^j. \tag{7.2.3-5}$$

Let $\#v_{5j}^Q$ be the number of correct digits in the coefficient v_{5j}^Q of the expansion for v^Q in (7.2.3-5) for all relevant values of $j \geq 0$. Here, correct means numerically correct. That is, the number of digits which are delivered by the computation as described following (7.2.3-4). Then let $\overline{\#v}_i$ be the average number of correct digits for all relevant values of $Q \geq 0$. This average is taken over every fifth coefficient in (7.2.3-5), and is defined as

$$\overline{\#v}_i = \begin{cases} \dfrac{1}{3}\sum_{j=0}^{2} \#v_{5j}^0, \\[2ex] \dfrac{1}{11}\sum_{j=0}^{10} \#v_{5j}^Q, & Q = 1, 2, \ldots . \end{cases}$$

The exceptional treatment for $Q = 0$ results from the fact that $v_j^F \equiv v^0 = 0$, $j > 10$. This in turn follows from the particular choice of γ_0 (cf. (7.2.3-1)) and from the definition of the relative rounding $S_{N_0}^{\gamma_0}$ (cf. (7.2.3-3), (5.4-10)f), and (3.1.2-43)).

In Fig. 7.2.3-2 we plot $\overline{\#v}_i$ versus a for $Q = 0, 1, 2, 3, 4, 5$. One additional curve $T(a)$ is plotted for comparison:

$$T(a) = -\log_{10}\| (I - S_{50}) Y_{exact} \|. \tag{7.2.3-6}$$

$T(a)$ is the number of digits in the remainder $\| (I - S_{50}) Y_{exact} \|$. No approximation to Y_{exact} on $S_{50}(\mathcal{M})$ can have a number of correct digits greater than this quantity. Thus $T(a)$ is an upper envelope for the family of curves $\overline{\#v}_i$ versus a. A second limiting factor is the floating-point mantissa length.

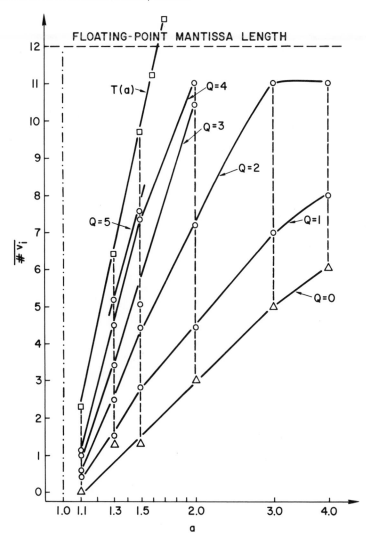

Figure 7.2.3-2

Thus, in fact

$$\overline{\#v_i}(a) \leq \max\ (T(a),12).$$

The curves in Fig. 7.2.3-2 show that our computations usually come within one correct digit of this theoretical bound. The steady corrective property of the IRC process itself is conveyed by the entire family of curves for $Q = 0, 1, 2, 3, 4, 5$.

2. We consider the example 2 of Section 7.2.1. The integral form of this example reads:

$$y = k(y) := y + \mathscr{L}y$$

$$= 16 \left(\frac{1}{8} - \int_0^{-1} \int_0^x w \, dx^2 - \int_0^1 \int_0^x w \, dx^2 \right)$$

$$+ x \left(\frac{1}{8} + \int_0^{-1} \int_0^x w \, dx^2 - \int_0^1 \int_0^x w \, dx^2 \right) + 8 \int_0^x \int_0^x w \, dx^2,$$

where

$$w = 2y + x + 1.$$

Then the residue of v is simply $r(v) = \mathscr{L}(v)$.

The computations are summarized in Table 7.2.3-1, where we see that the IRC process results in a gain of one or two digits (and up to 380% in relative error).

NONLINEAR EXAMPLES

In the nonlinear case the IRC process becomes more ramified, and we proceed now to describe the nonlinear version of IRC that we employ. (Of course, other versions are also possible.)

The fixed point equation

$$y = f(y) \tag{7.2.3-7}$$

is to be solved for y through IRC on the sequence of screens $S_N^{\gamma_n}(\mathcal{M})$, $n = 0, 1, 2, \ldots$. Let $\tilde{y}$ in the screen $S_N^{\gamma_n}(\mathcal{M})$ be a given approximation to y, where

$$\tilde{y} = S_N^{\gamma_n} f(\tilde{y}), \tag{7.2.3-8}$$

approximately. We seek a correction v to $\tilde{y}$ so that the given equation

x	$\lvert v^F(x) - v^L(x) \rvert$	$\dfrac{\lvert v^F(x) - v^L(x) \rvert}{\lvert v^L(x) \rvert}$
-1.0	0.046×10^{-12}	1.7×10^{-10}
-0.9	13.3×10^{-12}	3.8×10^{-10}
-0.8	16.1×10^{12}	2.2×10^{-10}
-0.7	15×10^{-12}	1.42×10^{-10}
-0.6	14×10^{-12}	1.0×10^{-10}
-0.5	11×10^{-12}	0.52×10^{-10}
-0.4	9×10^{-12}	0.42×10^{-10}
-0.3	7×10^{-12}	0.28×10^{-10}
-0.2	4×10^{-12}	0.14×10^{-10}
-0.1	2×10^{-12}	0.06×10^{-10}
0	0	0
0.1	2×10^{-12}	0.05×10^{-10}
0.2	4×10^{-12}	0.09×10^{-10}
0.3	7×10^{-12}	0.13×10^{-10}
0.4	8×10^{-12}	0.16×10^{-10}
0.5	11×10^{-12}	0.17×10^{-10}
0.6	13×10^{-12}	0.19×10^{-10}
0.7	16×10^{-12}	0.21×10^{-10}
0.8	16×10^{-12}	0.19×10^{-10}
0.9	13×10^{-12}	0.14×10^{-10}
1.0	0	0

Table 7.2.3-1

(7.2.3-7) is solved, viz.,

$$\widetilde{y} + v = f(\widetilde{y} + v). \qquad (7.2.3\text{-}9)$$

In the IRC process (7.2.3-9) is replaced by a screen-equivalent (i.e., a computer-implemented version), which we take to be

$$\widetilde{v} = S^{\gamma_m}_{\|\gamma_m\|} \, (f(\widetilde{y} + \widetilde{v}) - f(\widetilde{y}) + r), \qquad (7.2.3\text{-}10)$$

approximately, where

$$r = S^{\gamma_m}_{\|\gamma_m\|} \ (f(\tilde{y}) - \tilde{y}). \tag{7.2.3-11}$$

Typically $m = N + 1$. (7.2.3-10) itself is solved by iteration on the screen $S^{\gamma_m}_N(\mathcal{M})$,

$$\tilde{v}^{(j+1)} = S^{\gamma_m}_{\|\gamma_m\|} \ (f(\tilde{y}) + \tilde{v}^{(j)}) - f(\tilde{y}) + r). \tag{7.2.3-12}$$

In our computations $\|\gamma_m\| = 10$, $n = 0, 1, 2, \ldots$. Then let $\tilde{f}$ be the computer implemented version of f, and let

$$\tilde{y} = v_0 + v_1 + \ldots + v_{n-1},$$

the result of $n - 1$ steps of IRC. Then the iteration actually used for solving (7.2.3-10) for the nth correction v_n is conducted on the screen $(S^{\gamma_n}_{10}(\mathcal{M}); S^{\gamma_n}_{10}(\Omega))$ and is

$$v^{j+1}_n := \tilde{f}(v_0 + v_1 + \ldots + v_{n-1} + v^j_n)$$

$$\tag{7.2.3-13}$$

$$\boxminus \ \tilde{f}(v_0 + v_1 + \ldots + v_{n-1} \boxminus r_n), \quad j = 0, 1, \ldots \ .$$

Here $v^{(0)}_n = 0$ and

$$r_n := S^{\gamma_n}_{10}\Big(\tilde{f}(v_0 + v_1 + \ldots + v_{n-1}) - v_0 - v_1 - \ldots - v_{n-1}\Big).$$

Note that r_n is independent of the iteration over j. The iteration (7.2.3-12) is stopped when

$$\| v^{j+1}_n - v^j_n \| \le 10^{-12} \| v^j_n \|.$$

The nonlinear examples 3 and 4 now follow.

3. We consider the nonlinear boundary-value problem; Example 3 of Section 7.2.2. We plot the number of correct digits of the approximation $v^Q(x)$ versus x for $x \in [-1, 1]$. Q takes on the values $Q = 0, 1, 2, 3, 4$. In Fig. 7.2.3-3 we display the case $a = 1.5$, while the case $a = 2$ is shown in Fig. 7.2.3-4.

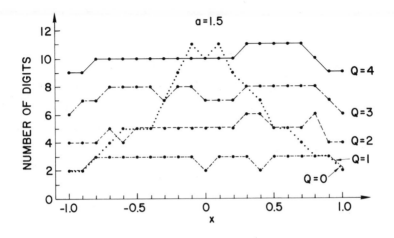

Figure 7.2.3-3

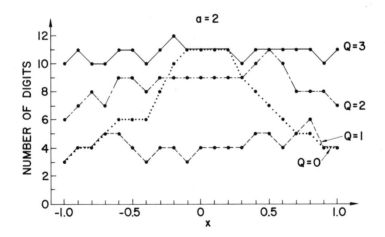

Figure 7.2.3-4

4. As a last example, we take the boundary-value problem

$$y' = 1 + x(y - x)^3,$$
$$y(1) - y(-1) - ay(0) = 1,$$

with the exact solution

$$y(x) = \frac{1}{\sqrt{(a^2 - x^2)}} + x.$$

We consider the case $a = 1.414213562373$ ($\sim\sqrt{2}$). In Figure 7.2.3-5 we plot the number of correct digits in the same manner as in the plots of the Example 3 just treated. In the present case only the values $Q = 0, 1, 2$ are plotted.

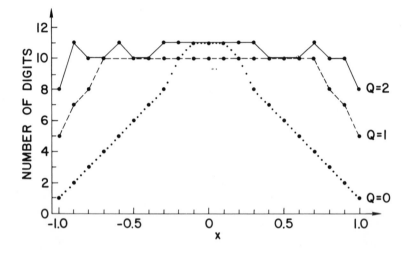

Figure 7.2.3-5

GLOSSARIES

There are two glossaries: a Glossary of Symbols and a Glossary of Concepts.

GLOSSARY OF SYMBOLS

The Glossary of Symbols is composed of three columns. The first column contains an alphabetized list of selected symbols which occur in the text. The second column contains a descriptive phrase for each such symbol. An entry in the second column may also contain a pointer ($\rightarrow$) to another entry in the glossary to which it is relevant. The third column identifies a chapter, section or subsection, as well as possibly an equation number where use or meaning of the glossary symbol is to be found.

i	isomorphism	3.1, 3.2
IR_N	isomorphic image (iIS_N)	3.2
$IR_N(P\mathcal{M})$	coefficient interval space	3.2
	parameter interval space	3.2
$I\!I\mathbb{R}[x](X)$	set of interval polynomials	3.2
$I\!I\mathcal{N}$	set of generalized intervals	3.2
$I\!I\mathbb{R}$	interval space	[1], [11]
IRC	Iterative Residual Correction	5.
IS_N	(directed projection), directed	3.2, (3.2-1)
	rounding, directed semimorphism	(3.2-5)
$IS_{N,i}$	i^{th} component of a spline rounding	(3.2.2-12)
$IS_N^{P'}$	relative rounding	(3.2.2-21)
$IS_N(P\mathcal{M})$	set screen of $P\mathcal{M}$	3.2
$IS_N(P\mathcal{M};\Omega)$	interval functoid: ($IS_N\ P\mathcal{M};IS_N\Omega$)	(3.2-6)
$\overset{\circ}{IS}_N(\Omega)$	$\{\overline{\oplus}, \overline{\ominus}, \overline{\odot}, \overline{\oslash}, \overline{\smallint}, \}$	(3.2-6)
$\overset{\circ}{\subset}$	*strict inclusion*	2.2, 3.2
$I\!I$ span Φ	interval subspace spanned by Φ	3.2

k	linear operator	(5.2-23)(5.3-3)
$\mathit{k}(\mathcal{M})$	space of sequences	3.1.1
$\mathcal{K}(Y)$	parametrized interval operator	2.2.1, (2.2.1-1),
		Theorem 4, (7.1.1-17), (7.1.3-5)

$\ell, \mathcal{L}$	linear operator	2.2, 4
$\lambda(A)$	*left bound* of interval A	3.1.2
$\lambda(\ell)$	eigenvalue of operator ℓ	2.2
$\Lambda(\ell)$	set of all eigenvalues (spectrum) of operator ℓ	2.2
LPB	Legendre Polynomial Basis	3.1.1, Table 3.1.1-1
$\mathcal{L}(Y)$	parametrized interval operator $\rightarrow \mathcal{K}(Y)$	2.2.1, Theorem 4

$\mathcal{M}$	function space	2.
MPB	Monomial Polynomial Basis	3.1.1, Table 3.1.1-1
M	$M \geq N$, dimension of a super screen	3.1, 3.1.2(i)
M'	$M' := M + 1$ degrees of freedom	3.1

$\mathcal{N}$	arbitrary Banach Space	3.2
$\tilde{\mathcal{N}}$	$\tilde{\mathcal{N}} \subset \mathcal{N}$	3.2
N	$N \in \mathbf{N}$, dimension of the screen	3.1
N'	$N' := N + 1$ degrees of freedom	3.1
$\mathbf{N}$	natural numbers	

Ω	$\Omega := \{+,-,\bullet,/,\smallint\}$	(3.1-9)
$\|\bullet\|_U$	norm with support U	2.2

GLOSSARY OF CONCEPTS

The Glossary of Concepts is a formal rearrangement of the Glossary of
Symbols. The first two columns are interchanged and the result is
realphabetized. Additional entries are made. The resulting glossary
may be used to locate a concept in the text based upon a descriptive
phrase for that concept. Pointers ($\rightarrow$) are used in the glossary to identify associated entries.

Bernstein Polynomial Basis	BPB	3.1.1, Table 3.1.1-1
Boundary conditions $\rightarrow$ constraints	$\mathscr{B}C$	(3.1.2-5)
backward Residual Correction	bwd RC	5.3
basis $\rightarrow$ sub-basis	$\Phi \rightarrow \Phi_N$	3.1, (3.1-1)
Chebyshev rounding		3.1.2, Table 3.1.1-1
Chebyshev Polynomial Basis	TPB	3.1.1, Table 3.1.1-1
constraints $\rightarrow$ boundary *conditions*	c	3.1.2(iii), 3.2.2(iii)
computer defined version of f	$\tilde{f}$	(7.1.1-3)
computer defined version of F	$\tilde{F}$	(7.1.1-9)
coefficient interval space	$IR_N(P\mathscr{M})$	3.2
coefficient space	$R_N(\mathscr{M})$	3.1
correct union	$\underline{\cup}$	(2.1-5)f
diameter of an interval A	$d(A)$	3.2
directed projection	IS_N	3.2, (3.2-1)
directed rounding	IS_N	3.2, (3.2-1)
directed semimorphism	IS_N	(3.2-5)
Exponential Monomial Basis	EMB	3.1.1, Table 3.1.1-1
eigenvalue of operator ℓ	$\lambda(\ell)$	2.2
existence		1., 2.1,
		Theorem 3-Theorem 7
		4., 7.
Identity operator	E	2.2.1, Theorem 1, (3.1.2-2)
Fourier Basis	FB	3.1.1, Table 3.1.1-1
forward Residue Correction	fwd RC	5.3
isomorphism	i	3.1, 3.2
Iterative Residual Correction	IRC	5.
function space	$\mathscr{M}$	2.
functoid	$(S_N(\mathscr{M}), \boxplus, \boxminus, \boxdot, \boxslash, \oint,)$	(2.1-7)

REFERENCES

[1]   Alefeld, G., and Herzberger, J.: *Introduction to Interval Analysis,* Academic Press, New York, 1983. Also appeared as *Einführung in die Intervallrechnung,* Reihe Informatik, Band 12, Wissenschaftsverlag des Bibliographischen Instituts Mannheim, 1974.

[2]   Bohlender, G., Böhm, H., Kaucher, E., Kirchner, R., Kulisch, U., Rump, S., Ullrich, Ch., and Wolff von Gudenberg, J.: *Wissenschaftiches Rechnen Programmiersprache,* (Kulisch, Ullrich Hrsg.) Berichte 10 des German Chapter of the ACM, Teubner, 1982.

[3]   Bohlender, G., Grüner, K., Kaucher, E., Klatte, R., Krämer, W., Kulisch, U., Miranker, W. L., Rump, S., Ullrich, Ch., and Wolff von Gudenberg, J.: PASCAL-SC: A Pascal for Contemporary Scientific Computation, IBM Report RC 9009 (No. 39456) (1981).

[4]   Bohlender, G., Kaucher, E., Klatte, R., Kulisch, U., Miranker, W. L., Ullrich, Ch., and Wolff von Gudenberg, J.: FORTRAN for Contemporary Numerical Computation, *Computing* **26,** 277-314, 1981.

[5]   Böhm, H.: Evaluation of Arithmetic Expressions with Maximum Accuracy, Proceedings of the IBM Symposium: *A New Approach to Scientific Computation,* (U. Kulisch and W. L. Miranker, eds.) Academic Press, New York, 1983.

254

[6]   Delves, L. M., and Freeman, T. L.: *Analysis of Global Expansion Methods: Weakly Asymptotically Diagonal Systems,* Academic Press, New York, 1981.

[7]   Epstein, C., Miranker, W. L., and Rivlin, T. J.: Ultra Arithmetic, Part 1: Function Data Types and Part 2: Intervals of Polynomials, *Mathematics and Computers in Simulation* **24,** 1-18, 1982.

[8]   Gottlieb, D., and Orszag, S. A.: *Numerical Analysis of Spectral Methods: Theory and Applications,* Soc. Ind. Appl. Math., 1977.

[9]   Kaucher, E., and Rump, S.: E-Methods for Fixed Point Equation $f(x) = x$, *Computing* **28,** 31-42, 1982.

[10]   Kaucher, E.: Solving Function Space Problems with Guaranteed, Close Bounds, Proceedings of the IBM Symposium: *A New Approach to Scientific Computation,* (U. Kulisch and W. L. Miranker, eds.) Academic Press, New York, 1983. Also appeared as Lösung von Funktionalgleichungen mit garantierten und genauen Schranken; im Berichte 10 des German Chapter of the ACM: Wissenschaftliches Rechnen Programmiersprachen, Teubner, 1982.

[11]   Kulisch, U. W., and Miranker, W. L.: *Computer Arithmetic in Theory and Practice,* Academic Press, New York, 1981.

[12]   Kulisch, U.: *Grundlagen des Numerischen Rechnens-Mathematische Begründung der Rechnerarithmetik,* Reihe Informatik, Band 19, Wissenschaftsverlag des Bibliographischen Instituts Mannheim, 1976.

[13]   Lanczos, C.: *Applied Analysis,* Prentice-Hall, Englewood Cliffs, 1956.

[14]   Rump, S.: Solving Algebraic Problems with High Accuracy, Proceedings of the IBM Symposium: *A New Approach to Scientific Computation,* (U. Kulisch and W. L. Miranker, eds.) Academic Press, New York, 1983.

[15]   Sadovskii, B. N.: Limit-Compact and Condensing Operators, *Russ. Math. Surveys* **27,** 1, 1972.

[16]   Stetter, H. J.: The Defect Correction Principle and Discretization Methods, *Num. Math.* **29,** 425-443, 1978.

[17]   Wolf R.: Aspekte des Nichtkompakhleitsmasses in der konstruktiven Funktionalanalysis, Überblicke Mathematik, 1977.